Linda K. Fuller, PhD

GW00708631

Yogurt, Yo
Yougho
An International Cookbook

*Pre-publication
REVIEWS,
COMMENTARIES,
EVALUATIONS . . .*

"**H**ere is a completely comprehensive presentation of everything yogurt. Linda Fuller has written a treatise which includes the history of yogurt, the nutritional benefits of yogurt, and yogurt recipes from around the world. If you have enjoyed a healthful, quick, and inexpensive lunch of yogurt-in-a-cup, wake up to the amazing array of yogurt recipes collected from over 50 different countries, all tested and enthusiastically approved. . . .

This is a book with emphasis on nutrition and good health, but it's the good taste you will enjoy."

Betsy W. Elliott
Author of *Bake 'n' Pack*
and *Love Chocolate*

Food Products Press
An Imprint of The Haworth Press, Inc.

Yogurt, Yoghurt, Youghourt
An International Cookbook

FOOD PRODUCTS PRESS

New, Recent, and Forthcoming Titles:

Nutrition Care of People with Diabetes Mellitus: A Nutrition Reference for Health Professionals by Penelope S. Easton, Charlotte S. Harker, Catherine E. Higgins, and Marvin C. Mengel

World Food and You by Nan Unklesbay

Yogurt, Yoghurt, Youghourt: An International Cookbook by Linda K. Fuller

Glossary of Vital Terms for the Home Gardener by Robert E. Gough

Yogurt, Yoghurt, Youghourt
An International Cookbook

Linda K. Fuller, PhD

Food Products Press
An Imprint of The Haworth Press, Inc.
New York • London • Norwood (Australia)

Published by

Food Products Press, an imprint of The Haworth Press, Inc., 10 Alice Street, Binghamton, NY
13904-1580

Library of Congress Cataloging-in-Publication Data

Fuller, Linda K.
        Yogurt, yoghurt, youghourt : an international cookbook / Linda K. Fuller.
            p.   cm.
        Includes index.
        ISBN 1-56022-034-1 (acid free).
        1. Cookery (Yogurt) 2. Cookery, International. I. Title.
TX759.5.Y63F84   1993
641.6'7146–dc20   92-45045
                                                                                                                        CIP

This book is dedicated to La Famille Rochat, the family with whom I lived in French-speaking Switzerland as a student on the Experiment in International Living. They introduced me not only to yogurt, but also to the joys of cultural sharing.

On a wider scale, it is dedicated to all who want to learn more about the world's foods and friendships.

ABOUT THE AUTHOR

Linda K. Fuller is a resident of Wilbraham, Massachusetts. She earned a BA in American Studies from Skidmore College, an MA in Human Technology from American International College, and a PhD in Communication Studies from the University of Massachusetts. Currently she is an Assistant Professor in the Media Department of Worcester (MA) State College.

In addition to serving as executive of the World Affairs Council, Linda has been very active on local community boards. She is the author of *Trips & Trivia: A Guide to Western Massachusetts* (Donning, 1988); *The Cosby Show: Audiences, Impact, Implications* (Greenwood Press, 1992); co-author (with Dr. Lilless McPherson Shilling) of *Communicating Comfortably, Communicating Quotably* and *Communicating About Communicable Diseases*, (HRD, 1990, 1993); and co-editor (with Paul Loukides) of the multi-volume *Beyond the Stars: Studies in American Popular Film* (Popular Press, 1990+).

For the Haworth Press, Linda is author of *Chocolate Fads, Folklore, and Fantasies: 1,000+ Chunks of Chocolate Information* (1994) and *Media-Mediated Relationships* (forthcoming). In addition, Dr. Fuller has been published in dozens of scholarly journals and has delivered more than 100 professional communications papers throughout the world.

Linda is married to Eric Fuller and they have three sons: William, Keith, and Alex–plus two Siamese cats and a *chocolate* labrador retriever named Truffles. A former tester, with her family, for Betty Crocker kitchens, she is famous locally as a cook of and an entertainer with ethnic dishes. The entire Fuller family–animals included–enjoy yogurt, no matter how it is prepared or spelled.

CONTENTS

Introduction

If you think yogurt tastes yucky, then this book will surely change your attitude. If you already think yogurt tastes yummy, then this book will be a great source of new recipes that are delicious, nutritious, low-calorie, and economical. All the recipes here uniformly call for one or two cups of plain yogurt, which can be purchased or homemade. There is nothing complicated.

The thrust of the book is international in scope, including more than 200 recipes from some 60 nationalities. Be sure to check out Appendix A on what yogurt (*Lactobacillus bulgaricus*) is called around the world, as well as Appendix B for a guide to the many countries represented here.

WHAT IS YOGURT?

Basically, yogurt is a curdled, cultured, semisolid food product, made from milk fermented by a bacterium. The resultant culture, which can be made from cow, goat, mare, reindeer, ewe, or water buffalo milk, has a slightly lower sugar content than pure milk, in addition to being a more easily digestible protein.[1]

Commercial dairies typically make yogurt by inoculating sterilized milk with the bacteria *Streptococcus thermophilus* or *Lactobacillus acidophillus*, then incubating the mixture several hours at 43° C (110-112° F), or until curd forms.

The World Health Organization (WHO) has developed standards of identity for the international use of yogurt, stating that it must be made with whole milk or partly skimmed milk, and it may contain milk powder, skimmed milk powder, unfermented buttermilk, whey (concentrated powder, proteins, or protein concentrate), water-soluble milk products, edible casein, and caseinate.

While the product is favored for persons with lactose intolerance, yogurt in the form of "bifidobacteria," a strain already prized for improving intestinal health by the Japanese and Europeans, will soon be produced in the United States.[2] Most yogurts have a shelf life of several weeks, if kept refrigerated, but various additives make a difference–and the same goes for freezing.[3]

A BRIEF HISTORY OF YOGURT

Yogurt dates back to pre-Biblical times, and Moses reputedly partook of it on his way to the Promised Land.[4] It has been estimated that man must have first learned about its properties during the Neolithic Era, once milk that was left to curdle had been sampled and found to be digestible. Yogurt undoubtedly made its premiere appearance in the Middle East, after the domestication of milk-producing animals began, around 9,000 B.C. The Great Temple at Ur contains friezes showing animals being milked. Genesis tells us that both Abraham and Moses served yogurt to their guests. There are also references to yogurt in the works of Herodotus, Homer, Pliny, Galen, and other ancient historians and physicians.

For four millenia, yogurt has survived and improved as a food product. *Larousse Gastronomique* recounts the following historical story about yogurt:

> King Francois I, suffering from an intestinal complaint which had resisted the whole "pharmacopoeia" of the day, heard that a Jewish doctor from Constantinople had been responsible for marvellous cures in similar cases, with milk curds prepared in a certain way. He brought this practitioner to Paris. The doctor arrived on foot with a flock of sheep and cured his royal client, but refused to divulge the secret of his concoctions.[5]

Yet the person most credited with elevating yogurt to its current status is the Russian scientist Élie Metchnikoff, co-winner (with Paul Ehrlich) of the 1908 Nobel Prize for physiology and medicine. Believing that longevity was influenced by digestion without putrefying bacteria, Metchnikoff studied the country of Bulgaria and was convinced that their use of cultured milk added to the population's impressive lifespans. In Metchnikoff's book *The Prolongation of Life: Studies of an Optimistic Philosophy*, he argued for the use of yogurt to prolong age. One of Metchnikoff's followers was a Spanish businessman, Isaac Carasso, whose son Daniel helped him in the manufacture and distribution of yogurt under the name "Dannon." Today, as evidenced in this book, yogurt is an international favorite.

WHO ENJOYS YOGURT?

Yogurt is popular around the world with people from all demographic niches.[6] It has been a staple in the diets of most Middle Eastern countries, notably Lebanon and Syria, as well as in Bulgaria, the former Yugoslavia, and India. Today, it is a familiar product in Europe, particularly Eastern Europe, and is beginning to gain a strong foothold in the United States, where it accounts for more than $800 million of the $20 billion dairy business.[7] Since antiquity, yogurt has been enjoyed by peasants and kings alike.

A survey conducted by the International Dairy Federation[8] found this configuration of industry-produced yogurt consumption around the world:

Country	Per capita yogurt consumption in kilos (1 kilo = 2.26 pounds)
Austria	2.9
Belgium	3.9
Canada	0.6
Great Britain	1.2
Israel	3.0
Finland	7.8
Netherlands	13.4
Switzerland	9.8
West Germany	4.7
USSR	0.1
USA	0.7

Almost universally, pediatricians are encouraging parents to introduce infants to yogurt at an early age because of its easy digestibility. At the same time that babies and toddlers are gobbling it up, the whole family is joining in. Yogurt's merit is not being ignored by senior citizens, either, who find it a particularly convenient and inexpensive product to have on hand. While we Americans currently consume more than a billion cups of yogurt each year,[9] that doesn't even take into account the frozen yogurt craze that is sweeping the country.

WHAT ARE YOGURT'S NUTRITIONAL PROPERTIES?

Around the world, yogurt is reputed to be a "miracle food" for dieters, babies, health nuts, or just folks who prefer natural foods. High in nutritional value, yet low in calories and fat, it has long been considered a perfect food–at the very least, an outstanding supplement to a well-balanced diet.[10]

This chart, from the U.S. Department of Agriculture,[11] provides critical nutritional information on the food properties of 100 grams (3 1/2 ounces) of yogurt:

Partially skimmed milk yogurt		Whole milk yogurt
Energy	50.0	62.0
Protein	3.4	3.0
Fat	1.7	3.4
Carbohydrates	5.2	4.9
Vitamin A	70.0	140.0
B vitamins		
Thiamine	.04	.03
Riboflavin	.18	.16
Niacin	.10	.10
Vitamin C	1.0	1.0
Minerals		
Calcium	120.0	111.0
Iron	trace	trace

With the growing press accounts on the dangers of osteoporosis, women in particular have come to appreciate yogurt as a low-calorie source of calcium.[12] Calorie-wise, here are some comparisons of various calcium sources:

1 cup	Calories
Milk	150
Lowfat milk	121
Lowfat yogurt	144
Skim milk	86
Sour cream	520
Yogurt	139

Although it is neither a panacea nor a cure-all, yogurt has been effective in treating certain cases of chronic constipation, dermatitis, ulcers, dysentery, vaginitis, and ileitis. Further, Tufts University has been involved in research in Finland using yogurt to help stop problems with diarrhea.[13]

There have also been curative claims about yogurt for ailments including something from nearly every letter of the alphabet: arthritis, bee stings, bursitis, constipation, dandruff, food poisoning, gall stones, halitosis, impotence, insomnia, kidney stones, migraines, Montezuma's Revenge, sunburns, ulcers, wrinkles, even vaginal infections.[14] And for those worried about cholesterol, it will come as a great joy to know that some studies indicate that yogurt might actually help lower cholesterol levels.[15]

SOME YOGURT FACTS AND FOLKLORE

• The country of Lebanon is named for "laban," the Lebanese term for yogurt.

• Because yogurt is predigested milk, the body gets nutrients out of it more quickly than it does from regular milk.

• In ancient Persia, a woman's dowry was determined in terms of the amount of "mast" (yogurt) her prospective husband could buy. Until that time, young girls were encouraged to include yogurt in their daily diet and as an ingredient in facials.

• Homer described yogurt repasts in his *Iliad* adventures.

• For facial beauty treatments, yogurt can double as both an astringent and exfoliant–just be sure to clean all of it off, since there is no truth to the notion that you can absorb its protein directly into your skin.

 • Genghis Khan was said to have fed yogurt to his armies when no other food was obtainable during their extensive marches through Mongolian and Persian Empires.

 • Truly viable yogurt has about 100,000 live bacteria to the gram.

 • Some of the most common preservatives added to yogurt include: citric acid, potassium sorbate, sodium benzoate, sodium citrate, and sorbic acid. It can be used as is, or sweetened by means of corn sweeteners, dextrose, honey, saccharin, sorbitol, sucrose, sugar, or sugar syrup. Yogurt's most frequent thickeners and stabilizers are agar, carob bean, carrageenan, corn sweeteners, kosher gelatin, locust bean gum, modified food starch, pectin, tapioca or tapolica starch, and vegetable stabilizer.

 • Mahatma Gandhi included an entire chapter on the value of yogurt in his book *Diet Reform*.[16] Indians use "dahi," plain or thinned with water, as a morning-after remedy; many also find yogurt to be a nerve-calmer and sleep-inducer.

 • "The Milk of Eternal Life," or *le lait de vie eternelle*, as it was called by Francis I of France,[17] was heartily touted by nutritionist Adelle Davis as a relaxant.

 • More than 90% of yogurt is digestible in one hour, compared to about 30% of milk.

THE STRUCTURE OF YOGURT, YOGHURT, YOUGHOURT

First, an explanation about this book's title. You obviously recognize the initial spelling, but might wonder about the others. In an effort to emphasize the international approach of this cookbook, I wanted to incorporate other appellations as well. The second spelling, "yoghurt," honors the fact that yogurt reportedly made its way to the United States in 1784 thanks to some Turkish immigrants.[18] Yoghurt was the Turkish word for the culture.[19] The third variation, "youghourt" is French, honoring my Swiss family mentioned in the Dedication, and about whom I shall refer again later.

Culinary sections that you will find in this book include these categories:

- Appetizers and Hors d'Oeuvres
- Soups and Salads
- Breads and Cakes
- Lunches, Suppers, Dinners
- Sweets for All
- Delicious Drinks

Recipes within each section are alphabetized, and the Index at the back is designed for easy cross-referencing. All ingredients are precisely specified, but the reader/cook is certainly encouraged to experiment.

Emphasis throughout *Yogurt, Yoghurt, Youghourt: An International Cookbook* is on the fact that yogurt–whether homemade or purchased–is delicious, nutritious, low-calorie, and economical.

MY OWN RELATIONSHIP WITH YOGURT

My initial encounter with yogurt came when I was living with a French-speaking family in Switzerland as an exchange student with the Experiment in International Living. We ate yogurt on grains for breakfast, mixed it into fascinating sauces for meals, sampled it with various fruits, stirred it into tantalizing pastries, whirred it up to make refreshing drinks, and had countless rewards with our many concoctions. I was hooked. This book is dedicated in warmest memory of and appreciation to La Famille Rochat, with whom I have kept in touch for more than a quarter of a century. Since my introduction to it, yogurt has been a part of my regular diet. My healthy husband and our three sons have profited enormously from our continuing association with yogurt.

Over the years, mainly through association with our local World Affairs Council, for whom I served as Executive Director for a dozen years, we have entertained numerous visitors from other countries. Many of them have shared their special recipes, allowing us to travel vicariously through ethnic eating.

In more recent years, I have had the opportunity to deliver a number of lectures in the media/communications field at the international level, and have been fortunate to have traveled to nearly all the countries listed in Appendix B. Whether I have enjoyed eating yogurt plain in the former Yugoslavia or Finland, in combinations in Singapore or Jamaica, used it for curative purposes in Egypt or Morocco, had it as a food supplement in Brazil or what was known as the Soviet Union, or heard stories about it from one of our sons when he was studying in Indonesia, yogurt has been an integral part of my life. This book is the result of the many wide horizons our family has gained from using yogurt.

ACKNOWLEDGEMENTS

At this point, I would like to state how pleasant it has been working with The Haworth Press, the publisher of my book *Chocolate Fads, Folklore, & Fantasies: 1,000+ Chunks of Chocolate Information* (1994) and *Media-Mediated Relationships* (forthcoming). It has been particularly pleasant working with Bob Gough, Senior Editor of Food Products Press, an imprint of The Haworth Press, as well as other Haworth staff who have worked on this project, including Bill Cohen, Bill Palmer, Eric Roland, Maryann O'Connell, Elizabeth Gould, Patricia Brown, and Lisa McGowan.

Thanks for this book also go to the following people: Betsy Elliott for her editing advice and rave reviews about the manuscript in its earliest stages; Judy Wilkinson for her cooking expertise; my niece Paige Harrigan for proofreading the manuscript; Sarah Lemly Knutson for early experiments with yogurt back in the Swiss days and a friendship that has lasted all these years; chef James Merrill for his willingness to experiment with many of these recipes; Nick Braithwaite for helping my son Will photocopy this in its rough stages; Donna Scott, Extension Associate with Cornell Cooperative Extension; as well as acknowledgements and appreciation for the many supportive friends who contributed recipes and reassurances; and, most of all, my family–who pre- and often re-tested many of these recipes, and who happily contributed to *Yogurt, Yoghurt, Youghourt.*

Linda K. Fuller
Wilbraham, Massachusetts

REFERENCE NOTES

1. Karen Cross Whyte, *The Complete Yogurt Cookbook* (New York: Ballantine, 1970), Introduction.

2. Barbara B. Deskins, PhD, "Dairy products for the nineties," *Cooking Light* (June 1992), 78.

3. "Naturally yoghurt" *GH* (March 1986), 180.

4. Sandra Lee Stewart, *Dannon Book of Yogurt* (Secaucus, NJ: Citadel Press, 1979), 17.

5. Prosper Montagne, *The New Larousse Gastronomique* (New York: Crown, 1977).

6. Judith Waldrop, "Do real men eat yogurt?" *American Demographics,* 13 (June 1991), 20.

7. Allison Otto, "The supermarket's $20 billion baby," *Dairy Foods,* 88 (January 1987), 24.

8. Carol Ann Rinzler, *The Signet Book of Yogurt* (New York: New American Library, 1979), 2-3.

9. Julee Rosso and Sheila Lukins, *The New Basics Cookbook* (New York: Workman, 1989), 452.

10. Sophie Kay, *Yogurt Cookery* (New York: Bantam Books, 1979), 2.

11. "Handbook #8," U.S. Department of Agriculture (Washington, D.C., 1976).

12. Yoplait has a little recipe booklet out called "Vive le calcium!" (Minneapolis: General Mills, 1986).

13. Jerry Bishop, "New form of yogurt helps stop diarrhea," *The Wall Street Journal* (September 5, 1991), B1.

14. "Yogurt as prophylaxis for candidal vaginitis," *American Family Physician* 46 (July 1992), 250 and "The yogurt-yeast infection connection," *Tufts University Diet & Nutrition Letter,* 10 no. 3 (May 1992), 1.

15. See Rebecca D. Williams, "Yogurt: The curds and whey to health?" *FDA Consumer,* 26, no. 5 (June 1992), 26 and Susan M. Kleiner, "Yogurt: A cultured cure-all?" *The Physician and Sportsmedicine,* 29, no. 4 (April 1992), 51-2 for more discussions on yogurt as a health resource.

16. Whyte, *Complete Yogurt,* xv.

17. Olga Smetinoff, *The Yogurt Cookbook* (New York: Pyramid, 1974), 15.

18. Kay Shaw Nelson, *Yogurt Cookery: Good and Gourmet* (Washington, DC: R.B. Luce, 1972), 13.

19. Sonia Uvezian, *The Book of Yogurt* (San Francisco: 101 Productions, 1981), 8. However, I must add that my friend Janee Friedmann, who lived in Turkey for several years, found the spelling of "yogurt" in her Turkish dictionary, which was also confirmed by the Turkish Embassy in Washington–who added that the "g" is silent when pronounced. Our cousin Helen Lovejoy Ernyey, currently serving in the U.S. Embassy at Ankara, agreed.

Appetizers and Hors d'Oeuvres

The beginning is the most important part of the work.

–Plato's *Republic*

Dazzle your family and friends from the start with the wondrous array of appetizers and hors d'oeuvres that can be prepared with yogurt.

When guests drop by unexpectedly, you can always present notable nibbles from items that you already have on hand, like *Aegean Anchovies*, which can be put together quickly.

If you want to be a little fancier, serve the *Appetizer Cheesecake, Italian-Style*, which blends ricotta cheese with antipasto-type vegetables and can be served either hot or cold, or prepare the impressive *Caspian Caviar Pie*, which is always a delight, or the fabulous *French Fruits de Mer*.

For dips, yogurt can do almost anything sour cream can do; yet, plain yogurt will be thinner and less viscous than sour cream, so it might need some kind of thickener. Remember, low fat yogurt has only 121 calories per cup, while sour cream has 520. As a sour cream substitute, you'll find yogurt adds a more tangy flavor to certain dishes.

The *Chiles Con Queso* dip is an outstanding opener to a Mexican dinner, as is the *Guatemalan Guacamole* or *Spanish Black Bean Dip*. *Crabby Hawaiian Dip* is positively sensational for summertime get-togethers, a special treat for a luau. And on Christmas Eve get into the tradition of serving *Finnish Pickled Herring*.

Egyptian Fava Beans, which makes two dozen fried cakes, is popular with young and old alike. It's always a good idea to have pita bread around, anyway. You can serve toasted pita bread or raw vegetables with the *Hummus*, a garbanzo bean/chickpea dip that is a Near Eastern specialty.

11

Or, make your own *Lebanese Yogurt Cheese*, which goes wonderfully with Arabic bread–serve plain with fruit, or combined with jams or herbs. Another choice might be *Liptauer Cheese*, a soft Hungarian combination of anchovies and capers also known as "Liptoi." The *Luscious German Liederkranz* recipe makes a favorite sandwich of our gourmet friend John Cooper, owner of "Sweet Daddy's" in Philadelphia.

Pâté no longer has to be a difficult undertaking. *Danish Liver Pate* (leverpostej) is easily assembled and can be served either as an appetizer or hors d'oeuvre. *Syrian Kibbe*, a baked meatloaf, and *German Party Meatballs* also fit into this same category.

If you want to have some fun, put out all the suggested condiments and let your guests enjoy *Jamaican Curried Eggs* as an appetizer; they'll have a fine time individualizing and experimenting with the hard-cooked eggs. Or, try *Polish Stuffed Eggs*.

Mouth-Watering Finnish Mushrooms (Paistetut Sienet) are fried, then served with a tangy yogurt sauce. *Cape Cod Clam Puffs* are easy to assemble.

Some splendid "skinny dips" can be made with yogurt, too–just stir in finely sliced and chopped vegetables, maybe adding a teaspoon or so of low fat powdered milk for bulk. Or, mix yogurt half-and-half with mayonnaise in your dips and spreads. Some of the Quick Dips included here are *Austrian Apricot-Peach Chutney Dip* and *Italian Anchovy Dip*. Let all your dips chill a few hours before serving, so the flavors can blend. *Romanian Eggplant Caviar* is an easy and inexpensive dip to have on hand.

A number of sauces are also included in this section, notably *Bombay Curry Sauce*, *Creole Sauce*, and *Dill-icious Swedish Sauce*. You can use them on their own or as accompaniments. The *Russian Caviar* makes an easy dish to bring as a housegift.

Shrimp is always a treat, and the *Japanese Gingered Shrimp*, made with a sake-based marinade, will earn you raves. So too will the *Pacific Salmon Spread*, which I like to serve in a fish-shaped mold.

Polynesian Pineapple is quite a conversation piece–a scooped out pineapple shell filled with a scrumptious yogurt concoction that is served with wheat crackers and celery sticks.

For the calorie- and calcium-conscious, the *Scandinavian Sar-*

dine Spread is a boon; as it says, three ounces of sardines contains 300 milligrams of calcium.

"Köttbullar," the *Swedish Meatballs*, double as either appetizers or entrees. I always make up lots, serving leftovers to my family for dinner. *Turquoise Yogurt* also has great adaptability–served as an appetizer, accompaniment, or as a hot or cold soup. *Gurkas Dilisas*, Norwegian cucumber appetizers, are always appreciated.

As you're beginning to see, yogurt has tremendous versatility. The last addition to this section, *Yummy Yogurt Combinations*, is merely meant to get you started.

APPETIZERS AND HORS D'OEUVRES

AEGEAN ANCHOVIES

These zesty little herring fillets date back to ancient Roman days.
Makes 1 1/2 cups.

2 whole canned anchovies (or 4 fillets), drained and chopped
1 Tablespoon anchovy paste
1 onion, chopped
1 cup plain yogurt
1/4 cup mayonnaise
1/4 teaspoon Worcestershire sauce

wheat crackers
capers

Mix together the anchovies, anchovy paste, onion, yogurt, mayonnaise, and Worcestershire sauce.

Spoon on individual thin wheat crackers and top each one with a caper.

APPETIZER CHEESECAKE, ITALIAN-STYLE

Ricotta cheese blends with antipasto-type vegetables.
Serves 4 to 6.

1 cup cheese crackers, crushed (18 to 20)
2 Tablespoons butter or margarine, melted
1 cup ricotta cheese
1 cup plain yogurt
4 eggs, slightly beaten
1 onion, chopped
2 stalks celery, chopped
6 to 8 stuffed green olives, minced
1 green pepper, seeded and chopped
1/4 cup parsley, chopped
6 Tablespoons flour
1 teaspoon dried mustard
1 teaspoon salt
1/4 teaspoon pepper
dash of hot pepper sauce

Topping:
    black olives, cut in rings
    bacon bits
    pimento strips

Preheat oven to 350°. Combine the cracker crumbs and melted butter, and use to line the bottom of a 9-inch spring-form pan.

Stir the ricotta cheese into the yogurt. Add the eggs, onion, celery, olives, green pepper, parsley, flour, mustard, salt, pepper, and hot pepper sauce. Pour into the cracker crust and bake until tested as done, about 45 to 60 minutes.

Remove, cover with toppings, and serve either hot or cold.

CAPE COD CLAM PUFFS

Served at the Kennedy Compound when I worked there for JFK.
Makes about 2 dozen.

2 cans (7 1/2 oz. each) minced clams, drained
2 3-oz. packages chive-flavored cream cheese, softened
2 onions, chopped
2 teaspoons Worchestershire sauce
1/4 teaspoon onion salt
dash of paprika
1 cup plain yogurt

melba toast rounds
Parmesan cheese, freshly grated

Add the clams to the softened cream cheese; blend in the onions, Worcestershire sauce, onion salt, paprika, and yogurt.

Pile the clam-yogurt mixture on top of melba toast rounds and sprinkle with freshly grated Parmesan cheese. Place under a broiler until the cheese is browned, 2 to 3 minutes.

CASPIAN CAVIAR PIE

The Caspian Sea region produces most of the world's caviar.
Serves 6 to 8.

1 envelope unflavored gelatin
2 Tablespoons lemon juice
4 oz. black caviar, preferably Russian
6 hard-cooked eggs, diced
1 cup plain yogurt
1 onion, finely chopped
1 teaspoon Worcestershire sauce
1/2 teaspoon salt
1/4 teaspoon pepper

cottage cheese, to garnish
rye crackers

Heat the gelatin in the lemon juice until it dissolves. Mix with the caviar, diced eggs, yogurt, onion, Worcestershire sauce, salt, and pepper. Transfer to a round 1-quart mold and chill until firm, several hours.

When ready to serve, unmold, put cottage cheese in the center, and serve with rye crackers.

CHILES CON QUESO

Serve this rich, spicy Mexican chile-cheese dip with crudites.
Makes about 3 cups.

1 onion, chopped
2 Tablespoons butter or margarine
1 cup solid-pack tomatoes, drained and chopped
1 can (4 oz.) green chiles, peeled, seeded, and chopped
1/2 teaspoon salt
dash of pepper
8 oz. Monterey Jack cheese, grated
1 cup plain yogurt

vegetable dippers or corn chips

Sauté the onion in butter until golden; stir in the tomatoes, chiles, salt, and pepper. Simmer, uncovered, about 15 minutes.

Stir in the cheese and yogurt and heat through. Serve in a chafing dish with vegetable dippers or corn chips.

CRABBY HAWAIIAN DIP

Lusciously impressive, but very easy to put together. This is an especially nice dip to have for a luau.
Makes 2 cups.

2 cans (6 oz. each) crabmeat, drained–or 1 lb. fresh crabmeat
8 oz. cream cheese, softened
1 cup plain yogurt
1/2 cup onion, chopped
1 Tablespoon parsley, chopped
1 Tablespoon Worcestershire sauce
1 teaspoon lemon or lime juice
1/2 cup macadamia nuts, chopped
dash of hot pepper sauce

chips or toast rounds

Add the crabmeat to the softened cream cheese, then blend together with the yogurt, onion, parsley, Worcestershire sauce, lemon or lime juice, macadamia nuts, and hot pepper sauce.

Serve with chips or toast rounds.

DANISH LIVER PÂTÉ

"Leverpostej" has many faithful devotees.
Serves 10.

4 Tablespoons butter or margarine
4 Tablespoons flour
1 cup plain yogurt
1 lb. beef liver, chopped
1/2 lb. salt pork, chopped
2 onions, chopped
1 egg
1 teaspoon onion salt
1/2 teaspoon pepper
cognac

toast rounds

Preheat oven to 350°. Melt the butter or margarine over low heat, then blend in the flour, stirring until smooth; gradually add the yogurt and cook, stirring constantly, until thickened. Remove the white sauce from the heat.

Combine the liver, pork, and onions and purée through a food grinder. Stir into the prepared white sauce and blend well. Add the egg, onion salt, and pepper, then put the mixture into an 8 x 4 inch loaf pan. Splash with some cognac, seal tightly with aluminum, and place this dish in a larger one. Prepare a hot water bath by pouring boiling water into the larger pan to a depth of about 2 inches.

Bake about 1 hour. Remove from the oven, take off the aluminum foil, and let cool. Store covered in the refrigerator until serving time, then cut into 1/2-inch slices. This versatile pâté serves 10 as an appetizer, or it can be served with toast rounds as an hors d'oeuvre.

EGYPTIAN FAVA BEANS

"Falafel" is considered to be Egypt's national dish.
Makes 2 dozen fried cakes.

1 can (19 oz.) fava beans, drained
1/2 cup cooked, mashed potato
1 onion, chopped
2 Tablespoons parsley, chopped
1 teaspoon turmeric
1/2 teaspoon coriander
1/2 teaspoon cumin
1/2 teaspoon onion salt
1/8 teaspoon cayenne pepper
2 eggs, slightly beaten
1/2 cup wheat germ

Sauce:
   1 cup plain yogurt
   1 teaspoon olive oil
   1 clove garlic, minced
   1 teaspoon dill, dried or chopped fresh
   1/2 teaspoon sesame seeds

pita bread

Purée the fava beans in a food processor; blend in the potato, onion, parsley, turmeric, coriander, cumin, onion salt, and cayenne pepper. Add the wheat germ to the slightly beaten egg, then combine with the fava bean mixture.

Shape the mixture into cakes about 2 inches in diameter and fry on both sides in olive oil until browned. Keep warm.

Serve in pita bread pockets with a sauce made of the yogurt, olive oil, garlic, dill, and sesame seeds.

FINNISH PICKLED HERRING

The Finns traditionally eat this on Christmas Eve.
Serves 12 family members and friends.

2 jars (6 oz. each) pickled herring
16 oz. can beets, drained and chopped
   (reserve 1 Tablespoon beet juice)
2 dill pickles, chopped
4 potatoes, peeled, boiled, and diced
2 apples, peeled and chopped
1 onion, chopped (optional)
salt and pepper to taste

Dressing:
   1 cup plain yogurt
   2 Tablespoons prepared mustard
   1/2 teaspoon sugar
   1 teaspoon cider vinegar
   1 Tablespoon beet juice

2 hard-cooked eggs, chopped

Drain the jars of herring fillets and chop coarsely. Combine with the beets, dill pickles, potatoes, apples, onion, salt, and pepper.

Make a dressing by combining the yogurt, prepared mustard, sugar, vinegar, and the reserved beet juice. Pour over the herring fillets, then sprinkle with the hard-cooked eggs as a garnish.

FRENCH FRUITS DE MER

The French call seafood "Fruits de Mer"; c'est magnifique!
Serves 4 to 6.

2 stalks celery, diced
1 green pepper, seeded and diced
1 Tablespoon oil
2 teaspoons Worcestershire sauce
1 cup plain yogurt
1 Tablespoon lemon juice
1 can (6 1/2 oz.) shrimp, drained
1 can (6 1/2 oz.) crabmeat, drained
1/2 teaspoon salt
1/4 teaspoon pepper
1 cup saltines, crushed (about 24)
1/2 cup Parmesan cheese, freshly grated

Saute the celery and green pepper in oil until tender. Stir in the Worcestershire sauce and yogurt. Drain the shrimp and crabmeat; combine with the lemon juice, salt, and pepper and add to the vegetables.

Put the seafood into a 1-quart, buttered casserole and top with crushed saltines and Parmesan cheese. Either bake at 350° about 20 minutes, or heat in a chafing dish until warm. Use as a dip or a spread.

Note: Leftover Fruits de Mer makes a marvelous topping on white fish–add during the last 5 minutes of baking.

GERMAN PARTY MEATBALLS

Serve these "Koenigsberger Klops" in a chafing dish.
Makes about 2 dozen.

2 lbs. equal parts ground beef, veal, and pork
1 onion, chopped
2 Tablespoons Worcestershire sauce
1 Tablespoon parsley, chopped
1 teaspoon paprika
1 teaspoon salt
1/4 teaspoon pepper
1/4 teaspoon lemon rind, grated
dash of nutmeg
2 eggs, slightly beaten
1/2 cup dried mustard

Sauce:
    1 can (10 1/2 oz.) onion soup
    1 Tablespoon cornstarch
    1 cup plain yogurt
    2 Tablespoons chopped pickles

2 Tablespoons parsley, chopped

Combine the ground meat, onion, Worcestershire sauce, parsley, paprika, salt, pepper, lemon rind, and nutmeg; stir in the eggs. Shape into meatballs about 1 inch in diameter. Roll in the mustard, then broil until browned.

Heat together the onion soup and cornstarch, then stir in the yogurt and chopped pickles. When ready to serve, top with the yogurt sauce and sprinkle with chopped parsley.

GUATEMALAN GUACAMOLE

This avocado spread is a specialty of Mayan/Mexican peoples.
Makes about 2 cups.

3 oz. cream cheese, softened
1 cup plain yogurt
2 ripe avocados, peeled and mashed
2 Tablespoons lime juice
1 clove garlic, minced
1/2 teaspoon onion salt
1/2 teaspoon chili powder
dash of hot pepper sauce

tortilla wedges or corn chips

Blend the yogurt into the softened cream cheese in a medium-sized bowl. Stir in the mashed avocado, lime juice, garlic, onion salt, chili powder, and hot pepper sauce; blend well.

Serve with tortilla wedges or corn chips for an authentic touch.

GURKAS DILISAS
(Norwegian Cucumber Appetizers)

These Norwegian cucumber tidbits can be made hours ahead.
Makes about 2 dozen hors d'oeuvres.

4 cucumbers
1 can (2 oz.) anchovy fillets, drained and chopped
2 packages (3 oz. each) cream cheese, softened at room temperature
4 green onions, minced
1 Tablespoon mayonnaise
1 teaspoon dill weed, dried or fresh, chopped fine
1/8 teaspoon cayenne pepper

melba toast rounds
1 cup plain yogurt
fresh chervil, to garnish

Wash the cucumbers but don't peel. Cut off the ends, scoop out the seeds, then slice into 1-inch chunks. Combine the chopped anchovy fillets with the cream cheese, green onions, mayonnaise, dill weed, and cayenne pepper; blend well. Stuff the mixture into the cucumber chunks, cover with plastic wrap, and chill several hours.

When ready to serve, put each cucumber piece on a toast round, top with a dollop of yogurt, and sprinkle with chervil. Serve cold.

HUMMUS

This garbanzo bean/chickpea dip is popular in the Middle East.
Makes 1 quart.

2 cans (20 oz. each) garbanzo beans/chickpeas, drained
1 cup plain yogurt
2 Tablespoons oil, preferably peanut
1/2 cup toasted sesame seeds
1/2 cup alfalfa sprouts
1 clove garlic, minced
1 teaspoon cumin
1/2 teaspoon salt
1/4 teaspoon cayenne pepper

Vegetable dippers or toasted pita bread

Drain and purée the garbanzo beans/chickpeas in a food processor. Blend in the yogurt, peanut oil, sesame seeds, alfalfa sprouts, garlic, cumin, salt, and cayenne pepper.

Chill until ready to serve, then use as a dip for fresh vegetables or toasted pita bread.

JAMAICAN CURRIED EGGS

Put the curry condiments on a lazy Susan for this appetizer.
Serves 6.

4 Tablespoons butter or margarine
2 onions, chopped
1 clove garlic, minced
1/2 teaspoon ground Jamaican ginger
3 Tablespoons flour
2 teaspoons curry powder*
2 cups coconut milk
1 Tablespoon lime juice
1 teaspoon salt
1/4 teaspoon pepper
12 hard-cooked eggs, halved

Condiments:
currants
mango chutney
cucumber cubes
1 cup plain yogurt
peanuts
banana slices

Melt the butter or margarine in a saucepan over low heat, and sauté the onions, garlic, and Jamaican ginger until the vegetables are softened. Blend in the flour and curry powder; cook, stirring constantly, about 5 minutes. Gradually add the coconut milk and lime juice and cook until thickened. Stir in the salt and pepper, then add the intact egg halves and heat through.

Serve with the following condiments: currants, mango chutney, cucumber cubes, 1 cup plain yogurt, peanuts, and banana slices.

Note: Here's my homemade curry powder recipe:

1/2 cup ground coriander
1/4 cup ground cumin:
1/2 teaspoon each:
ground pepper
black tumeric
black mustard seed
chili powder
salt

Measure all ingredients into a jar. Shake well and store in a cool place.

JAPANESE GINGERED SHRIMP

Sake, Japan's traditional rice wine, makes the marinade.
Serves 4 to 6.

2 lbs. shrimp, shelled, deveined, and cooked

Marinade:
    1/4 cup soy sauce
    3 oz. fresh ginger root, finely chopped
    1/4 cup white vinegar
    2 Tablespoons sugar
    2 Tablespoons sweet sake
    1 teaspoon salt

1 cup plain yogurt
3 Tablespoons green onions, chopped

Arrange the cooked shrimp in a shallow, oblong baking dish. Prepare the marinade: bring the soy sauce to a boil, add the ginger root, and simmer until most of the liquid is absorbed, about 5 minutes. Stir in the vinegar, sugar, sake, and salt. Pour over the shrimp, cover, and refrigerate several hours.

When ready to serve, drain and remove the marinated shrimp and serve with the yogurt mixed with green onions.

LEBANESE YOGURT CHEESE

"Lebanie" is best made with your own homemade yogurt.
Makes 2 cups.

4 cups plain yogurt*
1/2 teaspoon salt

Sprinkle the salt into the yogurt. Put several layers of cheesecloth in a colander, pour in the yogurt, and cover with a paper towel. Refrigerate and let drain for 24 hours.

Put the yogurt cheese into a container and refrigerate until ready to serve. Covered and refrigerated, it will keep up to 4 weeks.

"Lebanie" can be served plain or combined with jams or herbs and served with fruit or on Arabic bread.

*Note: If using store-bought yogurt, be sure it does not contain gelatin; otherwise, the straining process will not work.

LIPTAUER CHEESE

This soft, anchovy-flavored Hungarian treat is also known as "Lip-toi" cheese.
Makes about 1 quart.

2 cups small curd cottage cheese
4 anchovy fillets, chopped
2 teaspoons anchovy paste
1 teaspoon dry mustard
1 teaspoon Hungarian paprika
1/4 teaspoon salt
1 cup plain yogurt
1/2 cup butter, at room temperature
2 Tablespoons capers, drained and chopped
2 Tablespoons gin or schnapps (optional)

caraway seeds, to garnish
pumpernickel crackers or rounds

Combine the cottage cheese, anchovy fillets, anchovy paste, dry mustard, paprika, and salt in a blender; whirl until smooth. Add the yogurt, butter, and capers and blend together.

Stir in gin or schnapps. Put into a container and chill for several hours, preferably several days, before serving. Sprinkle with caraway seeds and serve with pumpernickel crackers or rounds.

MOUTH-WATERING FINNISH MUSHROOMS

"Paistetut Sienet" are Finland's fabulous fried mushrooms.
Makes 36.

3 dozen mushroom caps
2 eggs, beaten
1 teaspoon onion salt
1/4 teaspoon pepper
1/4 cup flour
1/4 cup cheese-flavored cracker crumbs
oil for deep-fat frying, heated to 350°

Yogurt sauce:
    1 cup plain yogurt
    1/4 cup mayonnaise
    1 Tablespoon capers, minced
    2 anchovy fillets, minced

Beat the eggs with the onion salt and pepper. Insert individual mushroom caps on a skewer; dip in the egg mixture, then the flour, and then the cracker crumbs. Let dry briefly, then deep fat fry until browned, 3 to 4 minutes.

Serve with a sauce made by combining the yogurt, mayonnaise, capers, and anchovies.

PACIFIC SALMON SPREAD

Fresh or canned, this recipe doubles as a canapé or a salad.
Serves 6.

3 oz. package lemon-flavored gelatin
1 cup boiling water
1 cup plain yogurt
1/4 cup shallots, minced
1/4 cup sweet pickles, minced
1/2 teaspoon salt
1/4 teaspoon pepper
1/4 teaspoon rosemary
1 lb. salmon, fresh (skinned, boned, and cooked) or canned

Topping:
    1 cup mayonnaise
    2 Tablespoons lemon juice
    parsley, capers, or sliced hard-cooked eggs, to garnish

Dissolve the gelatin in the boiling water and cool slightly. Stir in the yogurt, shallots, sweet pickles, salt, pepper, and rosemary; blend in the salmon.

Put into a 1-quart mold–I recommend a fish-shaped mold–and refrigerate until firm, about 3 to 4 hours. When ready to serve, top with the mayonnaise mixed with the lemon juice and choose an attractive garnish.

POLISH STUFFED EGGS

The Poles traditionally serve these at Eastertime.
Makes 48.

2 dozen eggs, hard-cooked and peeled
1 cup plain yogurt
4 Tablespoons parsley, chopped
2 Tablespoons chives, chopped
2 Tablespoons green onions, chopped
1/2 teaspoon salt
1/4 teaspoon pepper
paprika

Cut the eggs lengthwise in half; remove the yolks and mash them well. Mix with the yogurt, parsley, chives, green onions, salt, and pepper.

Stuff the egg halves with the yolk-yogurt mixture and sprinkle each one with paprika. Chill until ready to serve.

POLYNESIAN PINEAPPLE

A delicious and attractive conversation piece.
Serves 4 to 6.

1 whole pineapple, hollowed out
8 oz. cream cheese, softened
1 cup plain yogurt
1/2 cup shredded coconut
1/4 cup peach-pineapple chutney
1 teaspoon curry powder
toasted, slivered almonds

wheat crackers
celery sticks

Cut the hollowed-out pineapple sections into chunks; set aside.
Mix together the cream cheese, yogurt, coconut, chutney, and curry
powder. Combine with the pineapple chunks and chill.

When ready to serve, pile into the pineapple shell and top with
the slivered almonds. Serve with wheat crackers and celery sticks.

QUICK DIPS AND SAUCES WITH YOGURT

Austrian Apricot-Peach Chutney Dip
Makes 2 cups.

> 1 cup apricot-peach chutney
> *1 cup plain yogurt*
> 1 clove garlic, minced
> 1/2 teaspoon salt
> dash of pepper

Blend together the chutney and yogurt in a bowl; stir in the garlic, salt, and pepper. Chill until ready to serve.

Bombay Curry Sauce
Makes 1 1/2 cups.

> 2 Tablespoons butter or margarine, melted
> 1 Tablespoon green onions, minced
> 1 clove garlic, minced
> 1 Tablespoon parsley, chopped
> 2 teaspoons cumin seeds
> 1 teaspoon mustard seeds
> 1 teaspoon turmeric
> 1/4 teaspoon cardamon
> 1/4 teaspoon cayenne pepper
> *1 cup plain yogurt*
> salt to taste

Sauté the green onions and garlic in the melted butter. Add the parsley, cumin seeds, mustard seeds, turmeric, cardamon, and cayenne pepper; cook until the mustard seeds have popped. Cool. Blend in the yogurt and salt.

Creole Sauce
Makes 3 cups.

> 2 Tablespoons oil
> 2 onions, chopped
> 1 green pepper, seeded and diced
> 1 clove garlic, minced

1 can (15 1/2 oz.) stewed tomatoes
1/2 cup green olives, sliced
1 teaspoon salt
1/2 teaspoon sugar
1/4 teaspoon pepper
dash of hot pepper sauce
1 cup plain yogurt
1 Tablespoon sherry (optional)

Sauté the onion, green pepper, and garlic in the oil until tender, about 5 to 8 minutes. Add the stewed tomatoes, olives, salt, sugar, pepper, and hot pepper sauce; cook several minutes. Remove from heat and stir in the yogurt and sherry. Makes a nice sauce for shrimp.

Dill-icious Swedish Sauce
Makes 1 1/2 cups.

> *1 cup plain yogurt*
> 1/2 cup mayonnaise
> 1 teaspoon dill weed

zucchini slices
party rye bread

Combine the yogurt, mayonnaise, and dill weed; chill. Serve with zucchini slices either plain or on party rye bread.

Italian Anchovy Dip
Makes 2 cups.

> 1/2 cup butter
> 1/3 cup olive oil
> 2 oz. can anchovy fillets, drained and diced
> 1 clove garlic, minced
> 1/4 teaspoon oregano
> 1/4 teaspoon pepper
> *1 cup plain yogurt*

Italian breadsticks

Melt together the butter and olive oil; add the anchovies and garlic and sauté a few minutes until browned. Stir in the oregano, pepper, and yogurt. Serve with Italian breadsticks.

Luscious German Liederkranz
Makes 3 cups.

> 4 oz. Liederkranz cheese
> 1 cup ricotta cheese
> *1 cup plain yogurt*
> 2 Tablespoons lemon juice
> 2 Tablespoons chives, minced
> 1/2 teaspoon onion salt
> 1/4 teaspoon pepper
>
> pumpernickel bread rounds
> Bermuda onion slices

Soften the Liederkranz, then combine with the ricotta cheese, yogurt, lemon juice, chives, onion salt, and pepper. Serve on pumpernickle bread rounds, topped with Bermuda onion slices.

Russian Caviar
Makes 2 1/2 cups.

> 1 cup mayonnaise
> *1 cup plain yogurt*
> 6 Tablespoons chili sauce
> 6 Tablespoons caviar
> 2 Tablespoons pimentos, chopped
> 2 teaspoons chives, chopped
> 1/2 teaspoon parsley, chopped

Mix together the mayonnaise and yogurt in a bowl; stir in the chili sauce, caviar, pimentos, chives, and parsley. Chill.

ROMANIAN EGGPLANT CAVIAR

We had this the day we visited Dracula's castle in Transylvania.
Serves 6 to 8.

1 large eggplant, washed
1 cup onion, chopped
2 cloves garlic, minced
2 tomatoes, peeled and chopped
1 cup plain yogurt
2 Tablespoons olive oil
2 Tablespoons lemon juice, freshly squeezed
1/2 cup parsley, freshly chopped
1/2 teaspoon salt
dash of pepper

Preheat oven to 400°. Prick the eggplant in several places, place on an ungreased baking sheet, and cook until it is tender, about 45 minutes. Cool, then peel off the outer skin.

Put the eggplant pulp in a bowl and add the onion, garlic, tomatoes, yogurt, olive oil, lemon juice, parsley, salt, and pepper. Mix well and chill.

Before serving, drain off any excess liquid, then put into a dish and serve as an accompaniment to crudités and chunks of dark bread.

SCANDINAVIAN SARDINE SPREAD

3 oz. of sardines contain 300 milligrams of calcium!
Makes 2 cups.

3 (3/4 oz. each) cans bristling sardines, drained and mashed
8 oz. cream cheese
1 cup plain yogurt
4 Tablespoons parsley, chopped
2 Tablespoons lemon juice
1 onion, finely chopped (optional)
salt and pepper to taste

watercress, to garnish
crackers

Blend the sardines with the cream cheese and yogurt, preferably in an electric blender. Add the parsley, lemon juice, onion, salt, and pepper; mix well and put into a container. Refrigerate several hours or overnight.

When ready to serve, garnish with watercress and serve with crackers.

SPANISH BLACK BEAN DIP

The Spaniards call this "Salseo de Frijoles Negros," which our son
Will loved in Salamanca.
Makes 1 quart.

1 clove garlic
2 cups thick, cold, black bean soup
1 cup plain yogurt
2 Tablespoons mayonnaise
2 Tablespoons lemon juice
1 onion, finely chopped
1 Tablespoon olive oil
dash of hot pepper sauce
1 Tablespoon dry sherry (optional)

Rub the minced garlic into a serving bowl, then discard. Blend together the black bean soup, yogurt, mayonnaise, lemon juice, onion, olive oil, and hot pepper sauce. Stir in the sherry.

Cover and refrigerate, mixing again before serving. Serve with tortilla chips.

SWEDISH MEATBALLS

"Köttbullar" double as appetizers or entrees.
Serves 4 to 6.

2 lbs. ground beef and pork
2 eggs, slightly beaten
1 onion, chopped
1 cup bread crumbs
4 Tablespoons tomato paste
1/2 teaspoon salt
1/4 teaspoon thyme
dash of nutmeg
1/2 cup beef bouillon
1 cup plain yogurt
dill, to garnish

Slightly beat the eggs, then mix with the ground beef and pork, onion, breadcrumbs, tomato paste, salt, thyme, and nutmeg. Form into tiny meatballs about 1 inch in diameter. Cover with the bouillon and either bake at 350° about 20 minutes, or put in a slow-cooking pot for several hours or follow the manufacturer's instructions.

When ready to serve, stir in the yogurt and sprinkle lightly with dill.

SYRIAN KIBBE

"Kibbe bi-Sainieh" is a spicy baked meatloaf.
Makes about 48 appetizers.

1 lb. ground beef or lamb
1/2 cup bulgur
1 onion, chopped
1/2 teaspoon mint
1/2 teaspoon allspice
1/2 teaspoon salt
1/4 teaspoon pepper
1/2 cup pine nuts

Sauce:
   1 cup plain yogurt
   2 cloves garlic, minced
   1 Tablespoon fresh mint, chopped

Preheat oven to 350°.

Mix together the ground meat, bulgur, onion, mint, allspice, salt, and pepper; blend thoroughly. Put half the mixture into an 8-inch square baking pan, top with the pine nuts, and then cover with the rest of the meat.

Bake until firm and browned, about 30 to 35 minutes. Cut into about 48 squares.

Serve with a dipping sauce made by combining the yogurt, garlic, and mint.

TURQUOISE YOGURT

The French word for "Turkish," this has many uses.
Makes about 3 cups.

2 cucumbers, peeled, seeded, and diced
salt
1 clove garlic, minced
2 Tablespoons lemon juice
2 cups plain yogurt
1 teaspoon dill
2 Tablespoons olive oil
2 Tablespoons mint

Prepare the cucumbers, sprinkle with salt, and let stand for 15 minutes. Pour off the liquid and set aside.

Combine the garlic with the lemon juice in a bowl. Add the yogurt, dill, and cucumbers; mix together and chill. When ready to serve, top with olive oil and mint.

Turquoise yogurt is mostly used as an appetizer, served with toasted pita bread pieces or raw vegetables. It can also be an accompaniment to pilaf or served as a hot or cold soup.

YUMMY YOGURT COMBINATIONS

Add to *1 cup plain yogurt*, according to your taste:

> apples or applesauce
> apricots
> bananas
> blackberries
> cherries
> chocolate syrup
> grapefruit sections
> granola
> honey
> huckleberries
> lemon or lime juice
> maple syrup
> marmalade or other preserves
> nectarines
> oranges
> peaches
> pears
> pineapples
> prunes
> raspberries
> strawberries

(*Note:* This is only the beginning. Use your imagination . . .)

Soups and Salads

Yours is the Earth and everything that's in it.

–Kipling's *Rewards and Fairies*

The bounty of nature offers us endless delicacies that can be enhanced by yogurt.

Yogurt can be used as a substitute for milk, sweet cream, or sour cream in nearly any thickened meat or vegetable soup. From a quart of beef broth, *Armenian Bulgur-Yogurt Soup* and *Turkish Beef Soup* both blend up beautifully with a touch of mint. *Creole Callaloo* combines both vegetables and fruit, while *Dutch Pea Soup with Sausage* makes good use of a leftover ham bone.

Finnish Summer Soup (Kesakeitto) is worth looking forward to all year long garden vegetables provide the perfect accompaniment to a yogurt-flavored soup. Swedish *Apple-Cheddar Chowder* is fabulous in the Fall, and *Singaporean Soup with Mushrooms* is yummy year-round.

Seafood is a natural for yogurt. *Bisque de Cribiches*, a Caribbean crayfish/crawfish bisque, can be created from a base of either flavored water or fish stock, thickened with arrowroot. The combination of ingredients in *Haitian Avocado-Shrimp Soup* is bound to intrigue and entice you.

Any white fish can be used in *Scandinavian Fish Chowder Ménagerè*, a hearty soup that can serve as the entire meal. One of New England's specialties, in a tradition dating back to the days of the Pilgrims, is *Yankee Fish Chowder*.

You do not have to be Jewish to love and believe in the power of chicken soup. *Cambodian Chicken Soup* is a delightful specialty. And from "The Isle of Spices" comes a unique pepper-water soup: *Ceylonese Mulligatawny Soup*.

Fruit soups are particularly popular in warmer weather, but *Danish Cherry Soup* (Kirsebaersuppe) can be served either hot or cold;

the addition of claret or rose wine with Cherry Heering make this soup a real favorite. *French Blueberry Bisque* is sprinkled with cinnamon just before serving, and Pinot Chardonnay is added for that extra touch.

Probably the funniest soup inclusion in this book is *Hungarian Hangover Soup*, called "Tippler's Soup," which a Polish friend tells me actually does the trick. Is it the paprika, sauerkraut, sausage, spices, a combination of all, or perhaps the psychology?

"Corba," *Yugoslavian Lamb-Spinach-Rice Corba* soup, is a robust stew-like soup served at family get-togethers. As you can see, yogurt soups are happily served up in any number of ways.

Salads, too, favor yogurt. It can be substituted for cottage cheese in fruit salad, or used instead of sour cream as a base for chopped vegetables. *Afghani Eggplant-Yogurt Salad*, (Bonjan Borani) makes an amazing accompaniment to chicken or lamb, while *Cool Caribbean Salad* makes a marvelous side dish for ham.

Cucumbers are a frequent base for yogurt salads. *Albanian Cucumber-Yogurt Salad* is included here, plus several variations from other cultures.

"Bayerisches Kraut," *Bavarian Apple Slaw*, is a unique combination of red cabbage, apples, raisins, nuts, and other ingredients that blend beautifully with yogurt. Vegetables make up the base for *Iranian Potato-Pickle Salad* (Salad-e-Khiar Shur), including the protein of kidney beans.

Chilean Chicken-Corn Salad (Bocado Primavera de Ave) is a meal in itself, as is *Chinese Tuna Salad*. A delightful dinner salad, *Kashmir Curry of Lamb Salad* uses yogurt in its marinade.

Molded salads are yet another means for using yogurt; *Mousse au Jambon* will become one of your standard ways of using up leftover ham.

Just as yogurt goes so well with fish in soups, it also makes the ideal addition to salads. *Norwegian Summer Salmon Salad* (Sommer Salat) combines vegetables with salmon and is garnished with hard-cooked eggs and green beans. Broiled shrimp (Onigari Yaki) form the basis for *Shrimp-Egg-Pea Pod Salad, Japanese-Style*. And if you can get a hold of rock-lobster, be sure to treat yourself to *South African Rock-Lobster Mousse*. *Oriental Albacore-Bean Sprout Salad* blends protein with panache.

There are also lots of *Yogurt Salad Dressings* included here: *Arabic "Laban" Dressing, Balkan Boiled Dressing, Dill-icious Danish Dill Dressing for Tomatoes, Egyptian Creamy Cucumber (Salata-Zabady), French Yogurt Dressing, Green Goddess Dressing, Korean Fresh Fruit Salad Dressing, Low-Cal Lemon Dressing* (with lots of variations), *Mediterranean Orange Dressing, Oregano Italiano, Russian Dressing,* and *Turkish Tahini.*

SOUPS AND SALADS

AFGHANI EGGPLANT-YOGURT SALAD

"Bonjan Borani" is a great accompaniment to chicken or lamb.
Serves 4 to 6.

2 eggplants, about 1 lb. each
1 1/2 quarts boiling water
2 Tablespoons lemon juice
1 teaspoon salt

Dressing:
    3 Tablespoons lemon juice
    1/4 cup shallots, minced
    2 Tablespoons olive oil
    1/2 teaspoon salt
    1/4 teaspoon pepper
    1 clove garlic, minced
    1 cup plain yogurt

tomato wedges
fresh mint, chopped

Peel the eggplants and cut into 1-inch cubes. Cook in the boiling
water with lemon juice and salt until tender, 5 to 10 minutes. Drain
and cool, then chill.

Combine the ingredients for the dressing: lemon juice, shallots,
olive oil, salt, pepper, garlic, and yogurt. Pour over the chilled
eggplant cubes. Garnish with tomato wedges and mint before serving.

ALBANIAN CUCUMBER-YOGURT SALAD

Albanians call it "Kos Me Krastavec" and eat it often.
Serves 4.

2 cucumbers, peeled and cut into 1/2-inch cubes
1/2 teaspoon salt
1/4 teaspoon pepper
2 cloves garlic, minced
2 Tablespoons salad oil
1 cup plain yogurt

paprika, for garnish

Sprinkle the peeled, cubed cucumber chunks with salt and pepper; mix in the garlic. Stir the oil into the yogurt and combine with the cucumbers. Cover and refrigerate until chilled. When ready to serve, sprinkle with paprika.

Note: The Israelis add chopped walnuts and crushed mint to their cucumber-yogurt salad and call it "mastva khiar." The Sudanese include garlic powder in their "shorbat robe." Indian "raita" has freshly ground cumin, and Turkish "cacik" contains fresh dill. Arabic "khiyar bi-laban" doubles the amount of yogurt to cucumbers, and uses mint instead of paprika as a garnish.

APPLE-CHEDDAR CHOWDER

A Swedish specialty, "Appelsoppa" is served with rusks.
Serves 2 to 3.

1 can (10 1/2 oz.) cream of chicken soup
1 cup plain yogurt
1/2 cup milk or cream
1 cup cheddar cheese, grated
1/2 teaspoon brown sugar
1/4 teaspoon cinnamon
2 cups tart, green apples, peeled, seeded, and diced
1 onion, chopped (optional)

rusks
cinnamon

In a large saucepan, stir together the yogurt, milk, and cream of chicken soup. Add the grated cheese, brown sugar, cinnamon, apples, and onion.

Cook over low heat until the cheese melts. Garnish with cinnamon and serve with rusks.

ARMENIAN BULGUR-YOGURT SOUP

Spellings vary: Tanador, Tanabour, Tanabur.
Serves 4 to 6.

1/2 cup bulgur
water to cover
1 quart beef broth
1 teaspoon salt
2 Tablespoons olive oil
2 onions, chopped
2 Tablespoons parsley, chopped
1 Tablespoon fresh mint, chopped
2 cups plain yogurt
2 eggs, beaten

Soak the bulgur overnight in water to cover. Drain and put in a saucepan with beef broth and salt; cook until tender, about an hour.

Sauté the onions in the oil until soft, about 5 minutes, then add to the soup with the parsley and mint. Beat the eggs into the yogurt; slowly stir into the soup. Heat through but don't boil.

This unusual and healthy soup can be served hot or cold.

BAVARIAN APPLE SLAW

"Bayerisches Kraut" blends beautifully with apples.
Serves 4 to 6.

1 large head of red cabbage, shredded (about 4 cups)
2 cups tart apples, unpeeled but cored and diced
1/2 cup raisins
1/2 cup dill pickle slices
1 Spanish onion, thinly sliced
1/4 cup walnuts, chopped

Dressing:
   1 cup plain yogurt
   1/4 cup mayonnaise
   1/4 cup French dressing

Mix together the red cabbage, apples, raisins, dill pickle slices, onion, and walnuts in a large salad bowl.

For the dressing, stir the mayonnaise and French dressing into the yogurt. Pour dressing over the slaw and toss together.

BISQUE DE CRIBICHES

Caribbean crayfish/crawfish bisque.
Serves 6.

2 lbs. crayfish/crawfish, cleaned and shelled
4 Tablespoons butter or margarine
1 onion, chopped
2 cloves garlic, minced
1 whole fresh hot pepper (or a dash of hot pepper sauce)
fennel sprig
4 cups water or fish stock
1 cup plain yogurt
1 Tablespoon arrowroot
salt to taste

Melt the butter in a pan and add the crayfish, onion, and garlic; cook until the crayfish has changed color, about 5 minutes. Add the hot pepper, fennel, and water. Bring to a boil, then simmer 30 minutes.

Remove from heat, discard the hot pepper and fennel, and purée the crayfish before returning it to the stove. Add the yogurt. Blend a little bit of the stock into the arrowroot, then add to the soup and heat until thickened. When ready to serve, season with salt to taste.

Note: While crayfish is readily available in Martinique and Guadeloupe, the crawfish that is found in the United States works the same way for this recipe.

CAMBODIAN CHICKEN SOUP

"Moeun Sngo" is Southeast Asia's answer to the common cold.
Serves 6.

2 to 3 lbs. chicken, cut up
1/4 cup raw rice
2 scallions, chopped
2 cloves garlic, minced
1/2 teaspoon monosodium glutamate (MSG)
1 quart chicken stock
1 teaspoon sugar
1/4 cup lime juice
1 teaspoon salt
1/2 teaspoon pepper
1 cup plain yogurt

coriander, fresh or dried, for garnish

Put the chicken, rice, scallions, garlic, and MSG in the chicken stock. Bring to a boil, then simmer covered, until tender, about 30 minutes.

Remove the chicken from the soup. Discard bones and skin and cut the chicken into bite-sized serving pieces. Return to the broth and add sugar, lime juice, salt, and pepper; reheat. Just before serving, stir in the yogurt and sprinkle with coriander.

CEYLONESE MULLIGATAWNY SOUP

"The Isle of Spices" produces a unique pepper-water soup.
Serves 4 to 6.

2 Tablespoons butter or margarine
2 onions, chopped
4 carrots, peeled and thinly sliced
4 stalks celery, thinly sliced
2 tomatoes, peeled, seeded, and chopped
1 clove garlic, diced
2 Tablespoons flour
1 teaspoon curry powder
1/2 teaspoon coriander seeds
1/2 teaspoon cumin seeds
1/4 teaspoon ginger, fresh or dried
1 stick cinnamon
1 bay leaf
10 peppercorns
1 quart chicken stock
2 cups chicken, cooked and cubed
1 cup cooked rice
1 cup chick peas, cooked or canned
1 cup plain yogurt

lemon slices, for garnish

Lightly sauté the onions, carrots, and celery in the butter for about 5 minutes, then stir in the tomatoes, garlic, flour, curry, coriander seeds, cumin seeds, ginger, cinnamon stick, bay leaf, and peppercorns. Add the chicken stock and cook for about 2 hours, or until the vegetables are tender.

Strain the broth; add the cooked chicken, rice, chick peas, and yogurt. Heat through. Garnish with lemon slices.

CHILEAN CHICKEN-CORN SALAD

Called "Bocado Primavera de Ave," this is a hearty South American salad.
Serves 4 to 6.

4 cups cooked chicken, diced
1 can (1 lb.) sweet niblet corn, drained
4 tomatoes, peeled, seeded, and cut into wedges
2 green peppers, seeded and chopped
1 cup plain yogurt
1/2 cup mayonnaise
2 Tablespoons lemon juice
1/2 teaspoon chili powder
1/2 teaspoon salt
1/4 teaspoon pepper

Garnishes:
   hard-cooked eggs
   garbanzo beans
   black olives

Mix chicken, corn, tomatoes, and green peppers together. Combine the yogurt with the mayonnaise, lemon juice, chili powder, salt, and pepper; pour over the chicken-corn mixture and toss well.

Place on a lettuce-lined salad plate and garnish with hard-cooked eggs, garbanzo beans, and olives.

CHINESE TUNA SALAD

The Chinese call abalone "Bow Yu," a precious protein.
Serves 2 to 3.

1 can (7 1/2 oz.) white albacore tuna, drained
2 cups bean sprouts
1 can (7 1/2 oz.) water chestnuts, sliced and drained
1/2 cup white radishes, thinly sliced
1/2 cup cucumber, peeled, seeded, and chopped
1 green pepper, cut in slivers
1/2 cup green onions, sliced
lettuce greens

Dressing:
   2 Tablespoons soy sauce
   1 cup plain yogurt
   1/2 teaspoon celery salt

Garnishes:
   tomatoes
   hard-cooked eggs

Combine the white tuna, bean sprouts, water chestnuts, radishes, cucumber, green pepper, and green onions; put over lettuce greens.

Top with a dressing made from a mixture of the soy sauce, yogurt, and celery salt. Garnish with sliced tomatoes and hard-cooked eggs.

COOL CARIBBEAN SALAD

Fresh fruit abounds throughout the Caribbean islands.
Serves 4 to 6.

2 packages (3 oz. each) lemon-flavored gelatin
1 cup boiling water
1/2 teaspoon salt
1/2 cup pineapple juice
2 Tablespoons lemon juice
1 cup plain yogurt
1 can (13 1/2 oz.) crushed pineapple, drained (reserve juice)
1 cup melon balls or mango chunks
1 cup sliced strawberries
1 cup sliced peaches or papaya

mint, for garnish

Dissolve the lemon-flavored gelatin and salt in the boiling water. Add the pineapple juice that has been drained from the crushed pineapple, plus the lemon juice and yogurt. Freeze about 20 minutes, then beat with an electric beater about one minute, until fluffy.

Fold in the crushed pineapple, melon balls or mango chunks, strawberries, and peaches or papaya. Refrigerate in a 1-quart mold until set, about 2 to 3 hours. Garnish with fresh mint.

CREOLE CALLALOO

This West Indian soup is also spelled callilu, calaloo, calalou, or callau.
Serves 6.

3 lbs. callaloo (or spinach, Chinese spinach, or Swiss chard), mashed
12 okra
1 eggplant, peeled and chopped
1 quart water or vegetable stock
2 Tablespoons peanut oil
4 oz. salt pork, cut into 1/2-inch cubes
4 plantains or bananas, peeled and chopped
2 onions, chopped
2 cloves garlic, minced
2 Tablespoons chives, chopped
1 Tablespoon vinegar
1 teaspoon salt
1/2 teaspoon dried thyme
1/4 teaspoon ground cloves
1 fresh whole hot pepper (or a dash of hot pepper sauce)
1 cup plain yogurt

black pepper, for garnish

Put the callaloo leaves (or spinach, Chinese spinach, or Swiss chard), okra, and eggplant in the water; bring to a boil and cook until the vegetables are done, about 20 minutes.

Heat the oil in a pan and add the salt pork, bananas, onions, and garlic; cover and cook until tender. Add the chives, vinegar, salt, thyme, cloves, and hot pepper, and cook a few minutes more. Remove the salt pork cubes and put the rest of the soup through a sieve. Stir in the yogurt and heat through.

Sprinkle each portion with black pepper before serving. To make an authentic meal from Trinidad, Jamaica, Grenada, Haiti, Martinique, Guadeloupe, St. Lucia, or other Caribbean islands, serve with rice and codfish.

DANISH CHERRY SOUP

"Kirsebaersuppe" can be served either hot or cold.
Serves 6.

4 cups cherries (sweet or sour), pitted
2 cups water
1 cinnamon stick
2 whole cloves
4 thin slices of orange
4 thin slices of lemon
dash of salt
1 cup red wine (claret or rose)
2 Tablespoons Cherry Heering (optional)
1 Tablespoon cornstarch
2 Tablespoons water
1 cup plain yogurt

cinnamon, for garnish

In a saucepan, combine the cherries, water, cinnamon stick, cloves, orange and lemon slices, and salt. Simmer 10 minutes. Add the red wine and Cherry Heering.

Make a paste of the cornstarch and water, then add slowly to the soup; heat until thickened. Remove the cinnamon stick and fruit slices. At this point either stir in the yogurt and serve, or chill and use the yogurt as a garnish. Sprinkle with cinnamon just before serving.

DUTCH PEA SOUP WITH SAUSAGE

My kids call this "Sea Poop"; the Dutch call it "erwtensoep."
Serves 4 to 6.

3 cups chicken stock or broth
leftover ham bone
1 cup celery, chopped
1 onion, chopped
1 teaspoon onion salt
1/4 teaspoon black pepper
1/4 teaspoon marjoram
1 can (10 1/2 oz.) condensed green pea soup
1 cup plain yogurt
1 lb. smoked sausage, cooked and cubed

vodka (optional)

Garnishes:
   cooked peas
   parsley
   croutons

Bring the chicken stock, ham bone, celery, onion, onion salt, pepper, and marjoram to a boil, then simmer until ham bits fall off the bone. (Or, just heat the stock and seasonings, then add 1 cup of chopped ham.) Add the condensed pea soup, yogurt, and cooked sausage cubes; heat through.

Top with cooked peas, parsley, croutons, and about 1 Tablespoon vodka to consenting adults.

FINNISH SUMMER SOUP

During Finland's long summer days, "Kesakeitto" is savored.
Serves 8.

4 cups water or vegetable stock
2 teaspoons salt
1 teaspoon sugar
4 carrots, peeled and sliced
4 potatoes, peeled and cubed
1 medium head cauliflower, broken into florets
1 cup fresh green beans
1 cup freshly shelled peas
1 cup spinach, chopped
1/4 cup flour
1 cup plain yogurt
1 quart milk
2 Tablespoons butter or margarine

parsley, chopped, for garnish

Bring the water, salt, and sugar to a boil in a pan, then add the carrots, potatoes, cauliflower, and green beans. Simmer about 10 minutes, then add the peas and spinach and cook until the vegetables are tender-crisp.

Stir the flour into the yogurt and blend into the soup. Slowly add the milk and heat through. Stir in the butter and sprinkle with parsley before serving.

FRENCH BLUEBERRY BISQUE

Chilled berry bisque is a summertime treat "en français."
Serves 4.

1 pint blueberries, washed
2 cups water
1/2 cup sugar
1 cinnamon stick
1/2 teaspoon nutmeg
1 lemon, thinly sliced
1 cup plain yogurt
1/2 cup Pinot Chardonnay or other red wine

ground cinnamon, for garnish

Put the blueberries, water, sugar, cinnamon stick, nutmeg, and sliced lemon in a saucepan. Bring to a boil, then simmer 15 to 20 minutes. Remove the cinnamon stick and lemon slices.

At this point the blueberry soup can be strained or left intact. Chill, adding the yogurt and wine just before serving. Sprinkle each portion lightly with cinnamon.

HAITIAN AVOCADO-SHRIMP SOUP

The perfect combination for a midsummer night's eve . . .
Serves 4.

2 avocados, peeled, stoned, and mashed
1 quart chicken stock, slightly warmed
2 cups plain yogurt
1 Tablespoon lemon juice
1 teaspoon prepared mustard
1/2 teaspoon garlic salt
1 lb. shrimp, peeled, deveined, and cooked

1/4 cup parsley, chopped, for garnish

Blend the mashed avocados into the chicken stock, then slowly add the yogurt, lemon juice, mustard, garlic salt, and shrimp. Chill until ready to serve, then garnish with parsley.

HUNGARIAN HANGOVER SOUP

Called "tippler's soup," this sauerkraut/smoked sausage soup is the answer to antacids!
Serves 6 to 8.

2 onions, chopped
2 Tablespoons oil
1 Tablespoon Hungarian paprika
4 cups sauerkraut, drained and chopped
1 clove garlic, minced
1 lb. sausage (preferably Polish smoked), thinly sliced
1 teaspoon salt
1/4 teaspoon pepper
1 1/2 quarts water
2 Tablespoons flour
2 Tablespoons fresh dill weed, chopped
1 cup plain yogurt

croutons, for garnish

Heat the oil in a pan and brown the onions, about 5 minutes; stir in the paprika and sauerkraut. Add the garlic, sausage, salt, and pepper. Stir in the water, bring to a boil, and simmer 30 minutes.

Combine the flour, dill, and yogurt. Add slowly to the soup; heat through. When ready to serve, top with croutons.

IRANIAN POTATO-PICKLE SALAD

"Salad-e-Khiar Shur" includes an exciting mixture of great vegetables.
Serves 8 to 10.

4 potatoes, cooked and peeled
4 dill pickles, chopped
4 carrots, peeled, cooked, and sliced
1 can (1 lb.) red kidney beans, drained
12 radishes, sliced
4 green onions, chopped
1 small cabbage, shredded
2 Tablespoons fresh mint, chopped
1 teaspoon tarragon

Dressing:
    1/4 cup olive oil
    1/4 cup lemon juice
    1 clove garlic, minced
    1 cup plain yogurt
    2 Tablespoons prepared hot mustard
    1 teaspoon salt
    1/4 teaspoon pepper

Cube the cooked potatoes and put in a large bowl with the pickles, carrots, kidney beans, radishes, green onions, cabbage, mint, and tarragon.

For the dressing, combine the olive oil, lemon juice, and garlic; blend in the yogurt, prepared mustard, salt, and pepper. Pour over the vegetables and mix together. Chill, then toss again before serving.

KASHMIR CURRY OF LAMB SALAD

A delightfully different way to serve lamb.
Serves 4 to 6.

2 to 3 lbs. boneless lamb, cubed

Marinade:
  1/4 cup butter
  1 cup plain yogurt
  4 Tablespoons lemon juice
  1 onion, chopped
  2 cloves garlic, minced
  1 teaspoon coriander
  1 teaspoon salt
  1 teaspoon ground ginger
  1 teaspoon ground cardamon
  1/2 teaspoon cinnamon
  1/2 teaspoon black pepper

1 teaspoon curry powder

salad greens
chapatti bread
cucumber slices

For the marinade, melt the butter, then add the yogurt, lemon juice, onion, garlic, coriander, salt, ginger, cardamon, cinnamon, and pepper. Pour over the lamb cubes several hours, even overnight. Bake in a 350° oven 3 to 4 hours (or in a slow-cooking pot 8 to 10 hours), until lamb is tender.

When cooled, add the curry powder to taste. Pile on salad greens and serve with "chapatti" bread and cucumber slices.

MOUSSE AU JAMBON

A delightful French ham and lemon-flavored mold.
Fills a 1-quart mold.

3 oz. package lemon-flavored gelatin
2 cups cooked ham, diced
1 cup plain yogurt
4 stalks celery, finely chopped
1/4 cup shallots, finely chopped
1/4 cup green pepper, finely chopped
1/2 teaspoon dry mustard
1/4 teaspoon celery salt

Cook the lemon-flavored gelatin as directed on the package, adding only 1/2 cup cold water when the gelatin has dissolved. Chill until slightly firm, about 1/2 hour.

Add the ham, yogurt, celery, shallots, green pepper, dry mustard, and celery salt. Pour into a circular 1-quart mold and chill until firm, another 3 to 4 hours. Unmold and fill the center with more yogurt.

NORWEGIAN SUMMER SALMON SALAD

Called "Sommer Salat," but delicious year-round.
Serves 6 for lunch, 4 for dinner.

4 cups assorted salad greens, shredded
1 lb. smoked salmon, thinly sliced
8 to 10 mushrooms, sliced
1 cup beets, cooked and sliced
2 tomatoes, cut into wedges
6 radishes, sliced

Dressing:
  1 cup plain yogurt
  2 Tablespoons lemon juice or vinegar
  2 Tablespoons fresh parsley, chopped
  1 teaspoon prepared horseradish
  1 teaspoon fresh chervil, chopped
  1 teaspoon salt
  1/4 teaspoon pepper
  1/4 teaspoon basil

Garnishes:
  hard-cooked eggs
  cooked green beans

Put the salad greens in a salad bowl, then add the salmon, mushrooms, beets, tomatoes, and radishes.

For the dressing, mix together the yogurt, lemon juice or vinegar, parsley, horseradish, chervil, salt, pepper, and basil; pour over the salmon salad and toss. Garnish with hard-cooked eggs and cooked green beans, if desired.

ORIENTAL ALBACORE-BEAN SPROUT SALAD

High-protein and highly tasteful tuna blend beautifully with bean sprouts.
Serves 6.

6 albacore steaks, 1/2 lb. each
2 cups bean sprouts
12 white radishes, sliced
1 cucumber, peeled, seeded, and thinly sliced
2 scallions, chopped
2 carrots, peeled and thinly sliced

Dressing:
  1 cup plain yogurt
  1 Tablespoon soy sauce
  1/2 teaspoon onion salt
  dash of black pepper

lemon slices, for garnish

Cook the albacore steaks separately. Serve with the bean sprout salad: combine the bean sprouts, radishes, cucumber, scallions, and carrots.

Top with a dressing made from blending the yogurt, soy sauce, onion salt, and pepper. Garnish with lemon slices.

SCANDINAVIAN FISH CHOWDER MÉNAGERÈ

A robust soup, this is really a meal in itself.
Serves 6 to 8.

2 lbs. white fish (sole, bass, flounder)
1 Tablespoon butter
1 Tablespoon olive oil
2 onions, chopped
4 carrots, peeled and thinly sliced
4 potatoes, peeled and cut into small cubes
1 cup celery, thinly sliced
2 cups fish stock or water
1/2 teaspoon salt
1/4 teaspoon pepper
1 cup plain yogurt
1 egg yolk
1 teaspoon bouquet garni seasoning
1 teaspoon dill weed, chopped

melba toast rounds

Cut the fish into bite-sized chunks; set aside. Melt the butter and olive oil together and cook the onions until they are transparent, about 5 minutes. Add the carrots, potatoes, celery, fish stock, salt, and pepper. Bring to a boil and simmer until the vegetables are tender, about 30 minutes. Add the fish and cook until it is done, about 15 minutes.

Mix the yogurt with the egg yolk and bouquet garni, then slowly add to the soup. Heat, but don't allow it to come to a boil. Sprinkle with dill and ladle the chowder over melba toast rounds in individual bowls.

SHRIMP-EGG-PEA POD SALAD, JAPANESE-STYLE

Arrange the broiled shrimp "Onigari Yaki," picture-perfectly.
Serves 6.

1 cup rice, cooked
2 green onions, finely chopped
1 lb. shrimp, shelled, deveined, and broiled
4 hard-cooked eggs, divided
1/2 lb. pea pods, cooked

Dressing:
  1 cup plain yogurt
  1/2 cup French dressing
  1 Tablespoon soy sauce
  1 Tablespoon capers

salad greens
6 sweet gherkins, for decoration

Combine the cooked rice and chopped green onions. Add the shrimp, 2 of the hard-cooked eggs, diced, and the pea pods; chill.

Make a dressing of the yogurt, French dressing, soy sauce, and capers. Pour on top of the shrimp-rice mixture and serve over salad greens. Decorate the plate with the remaining hard-cooked eggs, sliced, and sweet gherkins.

SINGAPOREAN SOUP WITH MUSHROOMS

A speciality from the cleanest, safest country in the world.
Serves 4.

1 onion, chopped
2 Tablespoons peanut oil
1/2 pound fresh mushrooms, thinly sliced
1 Tablespoon lemon juice, freshly squeezed
1/8 teaspoon fresh nutmeg, grated
1/4 teaspoon salt
pepper to taste
2 cups chicken broth
1 cup plain yogurt

scallion sprigs, for garnish

Heat the oil in a medium-sized saucepan and sauté the onion until it is tender. Add the mushrooms and lemon juice and cook for about 3 more minutes, then stir in the remaining ingredients.

Warm the soup at a low temperature until heated through. Garnish with scallion sprigs, then enjoy this low-calorie, high-protein treat either as a first course or midday meal.

SOUTH AFRICAN ROCK-LOBSTER MOUSSE

These South African shellfish are shipped far and wide.
Serves 6.

1 envelope unflavored gelatin
2 Tablespoons water
1 to 2 cups rock-lobster meat, cooked
1/2 cup celery, finely minced
1/4 cup raw spinach, chopped
1 Tablespoon fresh parsley, chopped
1 teaspoon celery salt
dash of cayenne pepper
1 cup plain yogurt
1 Tablespoon lemon juice
1/2 teaspoon dry mustard
1/2 cup heavy cream, whipped

Filling suggestions:
    cottage cheese
    watercress
    sliced, marinated cucumbers
    diced tomatoes
    more yogurt, sprinkled with parsley

Sprinkle the unflavored gelatin over the water to dissolve. Mix together the lobster meat, celery, spinach, parsley, celery salt, and cayenne; set aside.

Combine the yogurt, lemon juice, and dry mustard. Fold into the softened gelatin, then add the lobster mixture. Whip the cream until stiff, then carefully fold in. Put the rock-lobster mousse into a 9-inch ring mold and refrigerate until firm, 3 to 4 hours. Unmold and fill the center with any of the suggested fillings.

SUMMERTIME SOVIET SALAD

Made from ingredients you already have on hand.
Serves 6 to 8.

1 small head of cabbage, shredded
4 cucumbers, peeled and thinly sliced
6 cooked potatoes, peeled and thinly sliced (preferably red)
6 cooked beets, peeled and sliced (or 1 lb. can, drained)
1 lb. parboiled fresh peas (or 10 oz. frozen package, thawed)
1 1/2 cups ham, cut into julienne strips

Dressing:
   3/4 cup virgin olive oil
   1/4 cup balsamic vinegar
   1/2 teaspoon salt
   dash of pepper
   herbs of your choice to taste

1 cup plain yogurt

Prepare and chill all the vegetables and the ham together. Blend the dressing ingredients, then mix and marinate everything except the yogurt for several hours. Just before serving stir in the yogurt.

TURKISH BEEF SOUP

"Yayla Chorbashi" is served at the May 1st festival of Hidrellz.
Serves 6.

2 lbs. stewing beef, cubed
2 Tablespoons oil
2 onions, chopped
1/4 cup wheat flour
1/2 cup pearl barley
1 quart beef stock
1 cup plain yogurt
2 Tablespoons fresh parsley, chopped
1 teaspoon dried mint
1 teaspoon salt
1/4 teaspoon pepper

Brown the stewing beef in a large kettle. In a separate sauce pan, sauté the onions in oil until transparent and tender, then stir in the wheat flour. Add to the browned beef with the barley and beef stock. Simmer about 1 1/2 hours, or until the beef and barley are tender.

Add the yogurt, parsley, mint, salt, and pepper and heat through. Sprinkle with some more parsley and serve hot.

YANKEE FISH CHOWDER

A tradition dating back to the Pilgrims.
Serves 6.

4 slices of bacon, cooked
2 onions, chopped
1 lb. white fish (halibut, cod, scrod), cubed
1 bottle (10 oz.) clam juice
2 cups potatoes, peeled and diced
1 teaspoon salt (less if you use cod)
1/4 teaspoon pepper
1 cup plain yogurt
1 1/2 cups milk

2 Tablespoons parsley, chopped
soda crackers

Fry the bacon; remove and crumble. Sauté the onions in the bacon grease until tender, then add the fish and cook through quickly, about 3 to 4 minutes. Add the clam juice, potatoes, salt, and pepper. Bring to a boil, then simmer until the potatoes are tender, about 15 minutes.

Stir in the yogurt, milk, and bacon bits. Heat through but do not boil. Sprinkle with parsley and top with soda crackers.

YOGURT SALAD DRESSINGS

Arabic "Laban" Dressing
Makes 1 1/4 cups.

> 2 cloves garlic, minced
> *1 cup plain yogurt*
> 1 teaspoon vegetable oil
> 1 teaspoon lemon juice or vinegar
> salt and pepper to taste

Blend the garlic into the yogurt, then add the oil, lemon juice or vinegar, salt, and pepper. Serve on salad greens or as a topping for fried vegetables.

Balkan Boiled Dressing
Makes 1 1/2 cups.

> 4 Tablespoons honey
> 2 Tablespoons flour
> 1 teaspoon salt
> 4 eggs, beaten
> *1 cup plain yogurt*
> 1/3 cup vinegar
> 1/2 teaspoon dry mustard
> 1/4 teaspoon pepper
> dash of cayenne pepper

Put the honey, flour, and salt in the top of a double boiler. Blend in the eggs, yogurt, vinegar, mustard, pepper, and cayenne. Cook over simmering water, stirring constantly, until thickened. This dressing is excellent on cabbage or cooked vegetables.

Dill-icious Danish Dill Dressing for Tomatoes
Makes 1 1/4 cups.

> 1 Tablespoon dill
> 2 green onions, chopped
> *1 cup plain yogurt*
> 1 teaspoon vinegar
> 1/2 teaspoon sugar

cucumber sticks, for garnish

Combine the dill and green onions and stir into the yogurt; stir in the vinegar and sugar. Serve either over tomato slices or one whole, opened tomato. Garnish with cucumber sticks.

Alternately, dill-icious Dill can also be slightly warmed and served over cooked broccoli, asparagus, spinach, brussels sprouts, green beans, etc.

Egyptian Creamy Cucumber (Salata-Zabady)
Makes 1 1/2 cups.

> 1 cucumber, peeled, seeded, and minced
> 1 teaspoon salt
> *1 cup plain yogurt*
> 1 Tablespoon salad oil
> dash of pepper
> 1 clove garlic, minced

Sprinkle the cucumber with salt and let stand 30 minutes; drain. Beat in the yogurt, oil, pepper, and garlic. Chill. Makes a nice accompaniment to a spicy meal.

French Yogurt Dressing
Makes 1 1/4 cups.

> 2 Tablespoons olive oil
> 2 Tablespoons lemon or lime juice
> *1 cup plain yogurt*
> 1/2 teaspoon salt
> dash of pepper

Blend the olive oil and lemon or lime juice into the yogurt in a bowl. Add salt and pepper, beat, and chill before serving.

For variation, add minced garlic, chopped shallots, herbs, paprika, or curry powder to taste.

Green Goddess Dressing
Makes 1 1/2 cups.

> 6 anchovies, finely chopped
> 1 green onion, chopped
> *1 cup plain yogurt*
> 2 Tablespoons mayonnaise
> 2 teaspoons tarragon vinegar
> 1 teaspoon lemon juice
> 1/4 cup fresh parsley, chopped
> 1/4 cup chives, minced

Combine the anchovies with the green onion in a bowl. Mix in the yogurt, mayonnaise, tarragon vinegar, lemon juice, parsley, and chives. Green Goddess goes especially well with San Francisco Bay shrimp.

Korean Fresh Fruit Salad Dressing
Makes 1 1/2 cups.

1 Tablespoon lime juice
1/4 cup brown sugar
1 cup plain yogurt
1/4 cup pine nuts, chopped

Stir together the lime juice and brown sugar and blend into the yogurt. Add the chopped pine nuts. Pour over fresh fruit: apple wedges, pear slices, pineapple chunks, melon balls, pitted cherries, orange sections, peach halves, etc.

Low-Cal Lemon Dressing
Makes 1 1/4 cups.

2 Tablespoons lemon juice
1 cup plain yogurt
1/2 teaspoon salt
1/4 teaspoon paprika
1/4 teaspoon dry mustard

Combine the lemon juice with the yogurt, salt, paprika, and mustard; chill before serving.
For variations:

a. Add 1/2 cup chopped onion and 1 minced garlic clove.
b. Substitute vinegar for the lemon juice.
c. Add 1/4 teaspoon basil or marjoram, 1 teaspoon Worcestershire sauce, 1 Tablespoon chili sauce, 1/2 teaspoon curry powder, or 1 teaspoon caraway or dill seeds.
d. For cheese flavor, add 4 Tablespoons cheddar cheese.
e. Add 4 Tablespoons blue cheese, 1 Tablespoon chives, and 2 teaspoons chopped parsley.
f. For fruit, substitute orange juice for the lemon and add 2 Tablespoons honey and 2 Tablespoons pineapple juice.

Mediterranean Orange Dressing
Makes 2 1/2 cups.

> 4 eggs, slightly beaten
> 1/2 cup honey
> 1/2 cup orange juice
> 2 teaspoons lemon juice
> 2 teaspoons orange rind, grated
> 1 teaspoon lemon rind, grated
> 1/4 teaspoon salt
> *1 cup plain yogurt*
> ground cinnamon

Beat the eggs, honey, orange and lemon juice together in the top of a double boiler. Stir in the orange and lemon rind and salt. Cook over simmering water, stirring constantly, until thickened. Remove, cover, and chill. Stir in the yogurt and sprinkle with cinnamon. Serve over salad greens or fresh fruit.

Oregano Italiano
Makes 1 1/2 cups.

> 1/4 cup honey
> 1 Tablespoon lemon juice
> *1 cup plain yogurt*
> 1 onion, chopped
> 2 Tablespoons parsley, chopped
> 1 teaspoon salt
> 1/2 teaspoon oregano

In a bowl, blend the honey and lemon juice into the yogurt, then add the onion, parsley, salt, and oregano. Oregano Italiano is terrific on tomatoes.

Russian Dressing
Makes 2 1/2 cups.

> 1 cup mayonnaise
> *1 cup plain yogurt*
> 1/3 cup chili sauce
> 4 scallions, chopped
> 2 Tablespoons lemon or lime juice
> 1/4 teaspoon salt

Stir the mayonnaise and yogurt together in a bowl. Add the chili sauce, scallions, lemon or lime juice, and salt. Serve with cold vegetables, shellfish salad, or on roast beef-rye bread sandwiches.

For variation, add chopped, cooked beets, horseradish, or caviar.

Turkish Tahini
Makes 1 1/2 cups.

> 4 Tablespoons sesame seeds
> 2 Tablespoons honey
> *1 cup plain yogurt*
> 2 cloves garlic, minced
> 1 teaspoon orange peel, grated

Toast the sesame seeds in a moderate oven (350°) until slightly browned, 3 to 4 minutes. Cool, then add to the honey, yogurt, garlic, and orange peel. Refrigerate several hours, then serve over fresh fruit.

YUGOSLAVIAN LAMB-SPINACH-RICE CORBA

"Corba" is a thick, hearty soup especially suitable for family get-togethers.
Serves 8.

2 lbs. ground lamb
2 onions, chopped
1 Tablespoon paprika
1 teaspoon salt
1/2 teaspoon pepper
8 cups water
1 cup rice, uncooked
10 oz. spinach (about 6 cups), washed, drained, and chopped
2 Tablespoons fresh parsley, chopped
1 cup plain yogurt
1 Tablespoon dill

Sauté the ground lamb and onions together in a saucepan until the meat is browned; blend in the paprika, salt, and pepper. Stir in the rice, spinach, and parsley. Add the water, bring to a boil, and simmer until the rice and spinach are cooked, about 15 minutes.

Combine the yogurt and dill and blend into the soup just before serving.

Breads and Cakes

Two things only the people anxiously desire–bread and circuses.

–Juvenal's *Satire*

Is there anything finer than the waft of baked goodies coming from your oven? You will have happy results using yogurt in your baking. When baking breads, it is usually best to have all the ingredients at room temperature, including the yogurt.

Once your guests have enjoyed your Mexican appetizers, serve up a batch of *Alamo Muffins, Sopa Paraguayn,* or *Spanish Rum Cake.* Other interesting ethnic breads you should try include *Far Eastern Flat Bread* (Naan), *Guinean Groundnut Bread,* which is made with peanut butter, *Swahili Soda Bread, Norwegian Walnut Bread,* and *Irish Soda Bread.*

The Swiss breakfast known as *Birchersmuesli* combines yogurt with rolled oats and any number of fruit, nut, or other toppings. *Health Bread* doubles for breakfast or teatime.

Apfelpfannkuchen (German apple pancakes) are a welcome treat for breakfast or simple Sunday suppers. Other alternatives include: *Belgian Beer Waffles* (Gaufres Bruxelloises), *Scandinavian Oatmeal Pancakes,* or *Swedish Pancakes* (Ugnspannkaka). And *Graham Griddlecakes* are a favorite in North America, served with pure maple syrup.

For your next teatime, *Blantyre Prune Teabread* is a delicious, nutritious Scottish treat. Be sure to try the *Scottish Scones* and *Trinidadian Orange-Coconut Bread. French Chocolate-Swirl Coffeecake* is nice served with café au lait.

When you make *German Cinnamon Coffeecake,* the recipe makes two coffeecakes–freeze one, if it doesn't get gobbled up. Try the *Israeli Coffeecake* for a pleasant change. Some popular additions from the U.S. are: *Great Lakes Granola-Raisin Coffeecake; Impeccable Pecan Muffins,* a Southern tradition, as is *Southern*

Spoon Bread; New England Blueberry Muffins; and *Cape Cod Cranberry Coffeecake*.

The summer squash known as zucchini, first developed in Italy, forms the basis for *Italian Zucchini Bread*. The wonderful molasses flavor in *New Zealand Double-Spice Muffins* is a native specialty.

You will soon find that yogurt greatly improves the flavor of your cakes, as well as your breads. As a general rule, when you use yogurt in baking, use 1/2 teaspoon of baking soda for each cup of yogurt.

Albanian Nut Cake is a wonderful accompaniment to Turkish coffee. Brazilian coffee is the choice to serve with *Brazilian Banana Cake*, which also uses native nuts, while *Columbian Cocoa Cake* (Pasteles de Cacao), of course, is served with Columbian coffee, and *Kerry Apple Cakes* are sumptuous with a steaming mug of Irish coffee.

For after dinner, *Cognac Cake à la Française* is a luscious treat with a snifter of brandy. Greek Brandy Cake (Yaourtini) tops off a great Greek dinner.

From the home of tortes, the most famous—*Austrian Linzertorte*—is included here, made with yogurt. Yet, it is not to be outdone by *German Chocolate Torte*, which serves 18 to 20, or *Hussar Torte*, the Hungarian chocolate-cherry creation.

Your favorite jam becomes the base for *British Blackberry Cake*, which makes quite a number of servings in its bundt pan.

Lemon is such a natural with fish that *Bulgarian Lemon Cake* is the perfect ending to any Black Sea fish dinner.

"Torta de Almendra," Spanish *Orange-Almond Cake*, has a funny story attached to it. Not only have lots of my friends volunteered recipes for this book, many were also willing to test them. Lynne Tower Combs, an old college friend who at the time was mayor of her town in New Jersey, made this cake for a group of women "who never talk recipes," and several of them privately asked her for copies!

Portuguese Cinnamon Cake (Bolo Pardo) is an Iberian specialty. Other sensational spices are featured in *Puerto Rican Spicecake* (Rizcocho de Especies).

The Swiss use native cherries for their *Swiss Cherry Cake* (Kirschenkuchen), but you can substitute canned cherry pie filling. Substitution is what this whole book is all about, anyway—using yogurt creatively, to suit your own individuality.

BREADS AND CAKES

ALAMO MUFFINS

A nice change from tacos or tortillas to accompany Mexican meals.
Makes 1 1/2 dozen.

1 1/2 cups yellow cornmeal
1 Tablespoon sugar
1 Tablespoon double-acting baking powder
1/2 teaspoon baking soda
1/2 teaspoon salt
1/2 cup shortening
2 eggs, slightly beaten
1 cup plain yogurt
1 can (8 3/4 oz.) cream-style corn
1/4 cup green chili peppers, seeded and minced

Preheat oven to 450°. Mix together the cornmeal, sugar, baking powder, baking soda, and salt. Using a pastry blender, cut in the shortening until the consistency is grainy. Beat the eggs into the yogurt, then add to the cornmeal mixture. Add the cream-style corn and chili peppers, mixing just until moistened.

Divide the batter evenly among 18 greased or paper-lined muffin cups. Bake until lightly browned, about 12 to 15 minutes.

ALBANIAN NUT CAKE

This nutritious nutty cake goes well with Turkish coffee.
Makes 24 servings.

1/2 cup butter or margarine
1 cup sugar
2 eggs, slightly beaten
2 cups flour
1 teaspoon double-acting baking powder
1 teaspoon baking soda
1/2 teaspoon cinnamon
1 cup plain yogurt
1 cup nuts (preferably walnuts), chopped
grated rind of 1 lemon

Glaze:
    1 cup water
    1 cup sugar
    1 teaspoon lemon juice

confectioner's sugar

Preheat oven to 350°. Cream the butter and sugar in a large mixing bowl; stir in the eggs. Sift the flour with the baking powder, baking soda, and cinnamon. Add the flour mixture and the yogurt alternately to the butter-sugar mixture. Stir in the chopped nuts and grated lemon rind, then pour the batter into a greased and floured 13 × 9 inch baking pan. Bake until a cake tester comes out clean, about 35 minutes.

While the cake is baking, make a glaze by boiling together the water, sugar, and lemon juice for 15 minutes. Poke some holes in the cooked cake, pour on the glaze, and return to the oven for 5 minutes. Remove and sprinkle with confectioner's sugar when cool.

APFELPFANNKUCHEN

These German apple pancakes are enormously exciting.
Makes 2 large pancakes.

4 Tablespoons butter or margarine, divided
2 apples, peeled, cored, and thinly sliced
4 eggs, beaten
1 cup flour
1 cup plain yogurt
1/2 teaspoon salt
1/4 cup sugar
1/4 teaspoon cinnamon
1/4 teaspoon allspice

fruit syrup
confectioner's sugar

Preheat oven to 400°. Put 2 9-inch round baking pans in the oven while it is preheating. Remove and add 2 Tablespoons of butter to each one; rotate around until butter is melted and the pans are coated. Divide the apple slices between the two pans.

Add the flour, yogurt, and salt to the eggs in a bowl; beat one minute. Pour the batter evenly on top of the apple slices. Top with a mixture of the sugar, cinnamon, and allspice. Bake uncovered until puffed and golden brown, about 20 to 25 minutes.

Top with fruit syrup and/or confectioner's sugar.

AUSTRIAN LINZERTORTE

From the home of tortes, the most famous of tortes.
Makes 9 servings.

1 1/2 cups flour, sifted
1/3 cup granulated sugar
1/2 teaspoon double-acting baking powder
1/2 teaspoon salt
1/2 teaspoon ground cinnamon
1/4 teaspoon baking soda
dash of ground cloves
1/2 cup butter
1/2 cup brown sugar
1 egg
1/3 cup almonds, ground

Cream filling:
   1 egg, slightly beaten
   1/3 cup sugar
   1/4 cup flour, sifted
   1/4 teaspoon salt
   1/2 cup milk, scalded
   1 cup plain yogurt
   1 teaspoon vanilla extract

1/2 cup raspberry jam

Preheat oven to 375°. In a large bowl, sift the flour with the sugar, baking powder, salt, cinnamon, baking soda, and cloves. Cut in the brown sugar and butter with a pastry fork until combined. Add the egg and ground almonds. Cover with waxed paper and refrigerate several hours, until well chilled.

For the cream filling: gradually beat the sugar into the egg and whip until thick and lemon-colored. Blend in the flour and salt. Add this mixture, along with the yogurt, to the scalded milk in the top of a double boiler. Cook this cream combination, stirring constantly, until thickened. Add the vanilla and cool.

With a spoon, press 3/4 of the dough onto the bottom of a 9-inch round layer-cake pan, with the sides slightly higher than the center.

Pour in the cream filling, then top with the raspberry jam. Roll out the rest of the dough to 1/4-inch thickness, and cut with a pastry wheel into 1/2-inch strips. Arrange the strips, lattice-fashion, over the raspberry jam; press to seal. Bake for 30 to 35 minutes or until lightly browned.

BELGIAN BEER WAFFLES

"Gaufres Bruxelloises" are light, crisp waffles.
Makes about 1 dozen.

3 1/2 cups flour, sifted
1/2 teaspoon salt
1/2 cup vegetable oil
1 1/2 pints light beer
2 eggs, slightly beaten
grated rind of 1 lemon
1 teaspoon fresh lemon juice
1 teaspoon vanilla extract

1 cup plain yogurt
1/3 cup brown sugar

In a large bowl, combine the flour, salt, oil, beer, eggs, lemon juice, lemon rind, and vanilla extract. Beat with a whisk until smooth, then let stand at room temperature about 2 hours to rise slightly–or refrigerate overnight. Bake according to your waffle iron directions.

To serve, top with the yogurt and brown sugar, either combined or served separately.

BIRCHERSMUESLI

It means "Swiss breakfast," but can be served anytime.
1 portion–individualized.

1 cup rolled oats
1 cup plain yogurt

Toppings:
  wheat germ
  nuts, chopped
  raisins
  seeds
  apples, grated
  berries
  honey
  brown sugar
  ice cream
  chocolate chips (!)

Combine the rolled oats and yogurt. Soak overnight in the refrigerator.

In the morning, add any combination of the suggested toppings, depending on your mood.

BLANTYRE PRUNE TEABREAD

A nutritious Scottish treat for teatime.
Makes 8 to 10 slices.

1 cup dried prunes, soaked, cooked, and chopped
2 cups white flour
1 cup wheat flour
1/3 cup sugar
2 teaspoons baking powder
1 teaspoon baking soda
1/2 teaspoon salt
2 eggs, slightly beaten
1 cup plain yogurt
1/4 cup butter or margarine, melted

Preheat oven to 350°. Prepare the prunes and set aside. Sift the flour with the sugar, baking powder, baking soda, and salt. Beat the eggs into the yogurt and add with the melted butter to the dry ingredients. Stir in the prunes.

Spoon the prune batter into a buttered 9 × 5 inch loaf pan and bake about 1 hour or until a cake tester comes out clean.

BRAZILIAN BANANA CAKE

Brazil combines Indian, Portuguese, and African cuisines.
Makes 8 servings.

1/3 cup butter or margarine
1 1/3 cups sugar
2 eggs
1 teaspoon vanilla extract
2 1/4 cups flour
2 teaspoons double-acting baking powder
1 teaspoon baking soda
1/2 teaspoon salt
1/4 teaspoon grated nutmeg
1 cup plain yogurt .
1 cup bananas, mashed (about 3)
1/2 cup Brazil nuts, chopped

whipped cream, sweetened, to accompany

Preheat oven to 350°. Cream the butter and sugar in a large mixing bowl. Add the eggs, one at a time, and beat well. Blend in the vanilla extract.

Sift the flour with the baking powder, baking soda, salt, and nutmeg. Add to the butter-sugar mixture alternately with the yogurt, beginning and ending with the flour mixture. Stir in the bananas and Brazil nuts. Blend, then pour into two greased and floured 8-inch round cake tins and bake until the cake tests done, about 35 minutes.

Remove from oven, let stand 10 minutes, then put on wire racks until cool. Serve as is, or frost between the layers and on the top with sweetened whipped cream. Serve with Brazilian coffee.

BRITISH BLACKBERRY CAKE

This spiced jam cake can be made with your favorite berry flavor.
Makes 16 servings.

1 cup butter or margarine
1 1/2 cups sugar
4 eggs
1 jar (10 oz.) blackberry jam
3 cups flour
1 teaspoon double-acting baking powder
1 teaspoon baking soda
1 teaspoon ground cinnamon
1/2 teaspoon ground allspice
1/2 teaspoon grated nutmeg
1/2 teaspoon salt
1 cup plain yogurt
1 cup nuts, chopped

confectioner's sugar, for topping

Preheat oven to 325°. Cream the butter and sugar in a large mixing bowl, then add the eggs one at a time; beat until lemon-colored. Stir in the blackberry jam.

Sift the flour with the baking powder, baking soda, cinnamon, allspice, nutmeg, and salt. Add the flour mixture and the yogurt alternately to the jam mixture. Fold in the chopped nuts.

Pour the batter into a greased and lightly floured bundt pan and bake until the cake tests done, about 1 1/2 hours. When cooled, sprinkle lightly with confectioner's sugar.

BULGARIAN LEMON CAKE

A summer delight, especially with Black Sea fish dinner.
Makes 12 to 16 servings.

1/2 cup butter or margarine
1 1/2 cups sugar
2 eggs
2 1/2 cups flour
2 teaspoons double-acting baking powder
1/2 teaspoon baking soda
1/4 teaspoon salt
1 cup plain yogurt
2 Tablespoons lemon juice
2 teaspoons lemon rind, grated

Glaze:
    4 Tablespoons lemon juice
    1 Tablespoon water
    1/2 cup sugar

Confectioner's sugar

Preheat oven to 350°. Cream the butter and sugar in a large mixing bowl. Add the eggs, one at a time.

Sift the flour with the baking powder, baking soda, and salt. Add the flour mixture and the yogurt alternately to the butter-sugar mixture, mixing at low speed with an electric mixer until well blended. Stir in the lemon juice and lemon rind, then pour into a greased and floured 13 × 9 inch baking pan and bake for 35 to 40 minutes.

When the cake has been out of the oven 5 minutes, puncture it in a few places and pour over a glaze made from the lemon juice, water, and sugar. Let stand until cool, then sprinkle with confectioner's sugar.

CAPE COD CRANBERRY COFFEECAKE

Serve at breakfast, or with the Big Bird at Thanksgiving.
Makes 12 to 16 servings.

1/2 cup butter or margarine
3/4 cup sugar
2 eggs
2 cups flour
1 teaspoon double-acting baking powder
1 teaspoon baking soda
1/2 teaspoon salt
1/4 teaspoon allspice
1 cup plain yogurt
1 teaspoon almond extract
1 can (7 oz.) whole-berry cranberry sauce

Frosting:
    1 cup confectioner's sugar
    2 to 3 Tablespoons milk
    1/2 teaspoon almond extract
    1/4 cup almonds, chopped

Preheat oven to 350°. Cream the butter and sugar in a large mixing bowl, then add the eggs one at a time. Sift the flour with the baking powder, baking soda, salt, and allspice; add to the butter-sugar mixture alternately with the yogurt. Stir in the almond extract.

Put half the mixture into a greased and floured bundt pan. Top with half the cranberry sauce; repeat this process. Bake until the cake tests done, about 1 hour. Cool for 5 to 10 minutes in the pan, then remove and cool completely.

Frost with a mixture of the confectioner's sugar, milk, and almond extract. Sprinkle with chopped almonds.

COGNAC CAKE À LA FRANÇAISE

The French do luscious little tricks with liqueurs.
Makes 24 servings.

1 cup butter or margarine
6 eggs
2 cups sugar
2 oz. cognac
2 teaspoons baking soda
1 cup almonds, finely ground
1 cup plain yogurt
2 cups flour
1 teaspoon ground cinnamon
1/2 teaspoon cloves

confectioner's sugar

Preheat oven to 350°. Beat the butter with the eggs and sugar in a bowl until light and lemon-colored. Dissolve the baking soda in the cognac, then add, with the ground almonds and yogurt, to the butter-sugar mixture.

Sift the flour with the cinnamon and cloves. Beat into the batter, then pour into a greased and floured 10 × 15 inch jelly-roll pan. Bake until the cake tests done, about 40 to 45 minutes.

Sprinkle with confectioner's sugar and use cookie cutters and your imagination to cut the cake into interesting shapes.

COLUMBIAN COCOA CAKE

Serve "Pasteles de Cacao" with Columbian coffee.
Makes 16 servings.

2 cups cake flour
1 cup sugar
1 teaspoon baking soda
1 teaspoon salt
1/4 cup cocoa powder
2/3 cup shortening
1 cup plain yogurt
1 teaspoon vanilla or peppermint flavoring
2 eggs

chocolate icing

Preheat oven to 350°. Sift the cake flour with the sugar, baking soda, salt, and cocoa; set aside. Blend the shortening with the yogurt and vanilla. Stir in the flour mixture and beat at medium speed with an electric mixer for 2 minutes. Add the eggs and beat 1 minute more. Pour into a greased and floured 13 × 9 inch cake pan and bake until the cake tests done, about 25 to 30 minutes.

Cool, frost with your favorite chocolate icing, and cut into serving-sized squares.

FAR EASTERN FLAT BREAD

"Naan" is a particularly popular bread in India and Pakistan.
Makes about 2 dozen small breads.

8 cups flour
1 cup plain yogurt
4 eggs, slightly beaten
2 Tablespoons double-acting baking powder
1 Tablespoon sugar
1 teaspoon salt
1/2 teaspoon baking soda
2 cups milk
vegetable oil
poppy seeds

In a large bowl, mix the flour, yogurt, eggs, baking powder, sugar, salt, and baking soda. Stir in enough milk to make a soft dough. Turn out onto a lightly floured board and knead the dough until smooth, about 5 minutes. Place in a greased bowl, then turn over. Cover with a towel and put in a warm, draft-free place for 3 hours.

Divide the dough in half, then keep dividing until you have about 24 pieces. Flatten each piece of dough on a lightly floured board and roll it into a 5-inch circle. Brush with vegetable oil, sprinkle with poppy seeds, and fry on a hot griddle on both sides, until golden brown and puffy.

FARINA CAKES

These Near Eastern dessert cakes are known as "Maamoul."
Makes about a dozen little cakes.

2 1/2 cups farina, cooked
1/2 cup butter, divided
1 cup plain yogurt
flour
2 cups pistachios, chopped
1 cup sugar
2 Tablespoons rose water (available at specialty stores)

confectioner's sugar

Preheat oven to 350°. Mix the cooked farina with 4 Tablespoons butter. Cover with the yogurt and refrigerate overnight, stirring occasionally.

When ready to bake, add the rest of the butter and enough flour to help the dough hold its shape. Make small cones out of the farina dough and fill with a mixture of the pistachios, sugar, and rosewater. Bake until golden brown, about 15 minutes. When ready to serve, sprinkle with confectioner's sugar.

FRENCH CHOCOLATE-SWIRL COFFEECAKE

Fantastique–especially when served with café au lait.
Makes 16 servings.

2 cups flour, sifted
1 cup granulated sugar
1 teaspoon double-acting baking powder
1 teaspoon baking soda
1 cup plain yogurt
2 eggs
1/4 cup butter or margarine

1 1/2 squares unsweetened baking chocolate, shaved and melted
1/2 cup walnuts, chopped
1/4 cup granulated sugar

Preheat oven to 350°. In a large bowl, beat together the flour, sugar, baking powder, baking soda, yogurt, eggs, and butter. Pour half the batter into a greased and slightly floured bundt pan. Combine the melted chocolate, nuts, and granulated sugar and drizzle half of it over the batter. Repeat this process.

Zigzag a knife through the batter to form chocolate-swirls. Bake for 45 to 50 minutes, or until the cake tests done. Sprinkle the top with more chopped nuts, if desired. Serve warm.

GERMAN CHOCOLATE TORTE

This impressive, layered German cake serves quite a gathering.
Makes 18 to 20 servings.

1 cup shortening
2 cups sugar
4 eggs, separated
1 teaspoon vanilla extract
4 oz. sweet German chocolate
1/2 cup water
2 1/2 cups flour
1 teaspoon baking soda
1/2 teaspoon salt
1 cup plain yogurt

Frosting:
    1/2 cup butter or margarine
    1 cup evaporated milk
    1 cup sugar
    2 egg yolks
    1 teaspoon vanilla
    1 can (3 1/2 oz.) coconut, flaked
    1 cup walnuts or pecans, chopped

Preheat oven to 350°. Cream the shortening and sugar together in a bowl, then add the egg yolks one at a time. Melt the chocolate and water together in the top of a double boiler, then add, with the vanilla, to the shortening-sugar mixture.

Sift the flour with the baking soda and salt, then add alternately with the yogurt to the batter. Beat the remaining egg whites until stiff, and carefully fold in. Divide the mixture between 3 greased and floured 9-inch round cake pans. Bake until the layers test done, 35 to 40 minutes.

Cool and cut each layer in half lengthwise, making 6 layers in all. Frost the layers and the top, but not the sides.

For the frosting: put the margarine, evaporated milk, sugar, egg yolks, and vanilla in a nonstick pan and cook, stirring constantly, until the mixture thickens–about 15 minutes. Add the coconut and nuts, then beat until spreadable.

GERMAN CINNAMON COFFEECAKE

Makes two–eat one and freeze the other.
Each coffeecake serves 9.

4 cups flour, sifted
1 Tablespoon double-acting baking powder
1 teaspoon baking soda
2 teaspoons grated nutmeg
1/4 teaspoon salt
1/2 cup butter
1/4 cup shortening
1 cup granulated sugar
4 eggs at room temperature
1 cup plain yogurt
1/2 cup milk
grated rind of 1/2 lemon

Topping:
    2 cups light brown sugar
    2/3 cup butter
    2 teaspoons ground cinnamon

confectioner's sugar

Preheat oven to 400°. Sift the flour with the baking powder, baking soda, nutmeg, and salt; set aside. In a large bowl, cream the butter and shortening with the sugar, then beat in the eggs, one at a time. To this creamed mixture add the flour mixture alternately in thirds with the yogurt and milk, beating gently each time. Stir in the grated lemon rind. The batter will be sticky and stiff.

Pour the mixture into 2 greased 9-inch square cake pans. Let the batter rest for 20 minutes, then cover with the *topping:* sprinkle 1 cup of light brown sugar on each of the coffeecakes, dot with butter, and generously top with cinnamon. Bake for 20 to 25 minutes or until the cakes test done. Let cool about 10 minutes, then dust with confectioner's sugar. Serve warm or cold.

GRAHAM GRIDDLECAKES

Griddlecakes are traditionally served in North America with pure maple syrup.
Makes about 2 dozen.

1 cup graham flour (or whole wheat)
1/2 cup white flour
2 to 4 Tablespoons sugar
1 teaspoon double-acting baking powder
1 teaspoon baking soda
1/2 teaspoon salt
1/4 teaspoon ground cinnamon
1 egg, beaten slightly
1 cup plain yogurt
1 cup milk
2 Tablespoons oil

butter
maple syrup (Vermont is our favorite!)

Sift the dry ingredients: flour, sugar, baking powder, baking soda, salt, and cinnamon. Separately beat the egg into the yogurt, milk, and oil.

Combine the mixtures until just moistened, then cook about 1/4 cup at a time on a preheated griddle (cast iron is best). Flip when puffed and browned on both sides, and serve with pats of butter and pure maple syrup.

GREAT LAKES GRANOLA-RAISIN COFFEECAKE

Grains from America's Great Lakes region make great granola.
Makes 9 generous servings.

1 cup white flour
1/2 cup whole wheat flour
1/2 cup brown sugar or honey
2 1/2 Tablespoons double-acting baking powder
1 teaspoon baking soda
1/2 teaspoon salt
2 eggs, slightly beaten
1 cup plain yogurt
1/4 cup safflower oil
1 teaspoon vanilla extract
1/2 cup raisins

Topping:
    1 cup granola
    1/4 cup brown sugar
    1 teaspoon ground cinnamon
    2 Tablespoons butter

Preheat oven to 350°. Blend the flour with the brown sugar or honey, baking powder, baking soda, and salt. Beat the eggs and add the yogurt, oil, and vanilla; fold into the flour mixture. Add the raisins.

Pour into a greased 8-inch square baking pan. Top with a mixture of the granola, brown sugar, and cinnamon. Dot with the butter and bake until the cake tests done, about 30 to 35 minutes.

GREEK BRANDY CAKE

"Yaourtini" is an elegant addition to a Greek feast.
Makes 9 servings.

1/2 cup butter
1 cup sugar
2 eggs
1 3/4 cups flour
2 teaspoons double-acting baking powder
1/4 teaspoon salt
1 cup plain yogurt
1 teaspoon baking soda
1 Tablespoon brandy

Decoration:
   confectioner's sugar
   almonds, slivered
   candied cherries, sliced

Preheat oven to 375°. Cream together the butter and sugar at medium speed using an electric mixer. Lower the speed and beat in the eggs one at a time until lemon-colored.

Sift the flour with the baking powder and salt. Mix the baking soda into the yogurt. Then add the flour mixture and the yogurt alternately to the batter. Pour in the brandy, then beat at high speed 1 to 2 minutes. Turn into a buttered 9-inch square cake pan, put into the oven, and immediately lower the heat to 350°. Bake until the cake is a light, golden color and springs back to the touch, about 35 to 40 minutes.

Cool the cake for a few minutes on a wire rack, then dust with confectioner's sugar. When fully cool, dust again with the sugar and decorate with the slivered almonds and candied cherries.

GUINEAN GROUNDNUT BREAD

Africans use lots of groundnuts (peanuts) in their cooking.
Makes 6 to 8 servings.

1/2 cup peanut butter, preferably chunky
1/4 cup sugar
1 egg, slightly beaten
2 Tablespoons butter or margarine, melted
1 cup plain yogurt
2 cups flour
1 teaspoon double-acting baking powder
1 teaspoon baking soda
1/2 teaspoon salt
1/3 cup peanuts, chopped

Preheat oven to 350°. In a large mixing bowl, combine the peanut butter, sugar, and egg. Stir in the melted butter and yogurt.

Sift the flour with the baking powder, baking soda, and salt. Combine with the peanut butter mixture and turn into a 9 × 5 inch buttered loaf pan. Sprinkle with the chopped peanuts and bake until the bread tests done, about 1 hour.

HEALTH BREAD

In any country, a nice gift to salute "your health!"
Makes 6 to 8 servings.

1 cup white flour
2 cups graham flour
1/2 cup brown sugar
1 teaspoon salt
1 teaspoon baking soda
1 cup plain yogurt
1/2 cup milk
1/2 cup molasses

Preheat oven to 325°. Sift the flour with the sugar and salt into a large bowl. Dissolve the baking soda in a mixture of the yogurt and milk, then add, with the molasses, to the dry ingredients.

Pour into a greased 9 × 5 inch loaf pan and bake until the bread tests done, about 1 1/2 hours. Toasted and buttered, this bread is in a category all its own.

Note: I won a baking contest with this recipe!

HUSSAR TORTE

Hungarian chocolate-cherry "huszar torte" is a gâteau flavored with cherry liqueur.
Makes about 16 servings.

2 1/2 cups cake flour, sifted
1 1/2 teaspoons double-acting baking powder
1/2 teaspoon baking soda
3/4 cup butter or margarine
1 cup granulated sugar
3 eggs
1/2 cup cherry or currant jam
1/2 teaspoon ground cinnamon
1/4 teaspoon ground nutmeg
1/8 teaspoon ground cloves
1 square unsweetened baking chocolate, grated
1 cup plain yogurt
1/4 cup cherry liqueur
1 cup heavy cream
2 Tablespoons sugar
1 teaspoon vanilla

Decoration:
   chocolate curls
   cherries

Preheat oven to 350°. Sift the flour with the baking powder and baking soda; set aside. In a large bowl, cream the butter and sugar with an electric mixer at medium speed, then add the eggs, one at a time; beat until fluffy. Add the cherry or currant jam, cinnamon, nutmeg, cloves, and grated chocolate. To the batter, add the yogurt alternately with the flour mixture, blend, and then turn into a greased and floured bundt pan.

Bake for 50 to 60 minutes, or until a cake tester comes out clean. Cool 10 minutes in the pan, then turn out onto a wire rack and cool completely.

When cool, cut the cake in half horizontally and sprinkle the cut surfaces with cherry liqueur. Whip the cream until stiff, adding the sugar and vanilla, and spread between the layers and over the top. Decorate with chocolate curls and cherries.

IMPECCABLE PECAN MUFFINS

Pecans are associated with American Southern cooking and traditions.

Makes 1 dozen.

1 cup white flour
1 cup whole wheat flour
4 Tablespoons brown sugar
1 Tablespoon double-acting baking powder
1/2 teaspoon salt
1/4 teaspoon baking soda
1 cup pecans, chopped
1 cup plain yogurt
4 Tablespoons butter or margarine, melted
2 eggs, beaten

granulated sugar, as a topping

Preheat oven to 375°. Sift the flour with the brown sugar, baking powder, salt, and baking soda. Add the chopped pecans. Melt the butter and blend into the yogurt; add the eggs. Combine the dry ingredients and the yogurt just enough to moisten.

Divide the batter evenly among 12 greased or paper-lined muffin cups. Sprinkle lightly with sugar and bake until done, about 15 to 18 minutes.

IRISH SODA BREAD

A standard for St. Patrick's Day!
Makes 2 loaves, 6 to 8 servings each.

2 cups white flour
2 cups whole wheat flour
1/2 cup sugar
1 teaspoon baking soda
1 teaspoon salt
1/2 teaspoon ground cardamon or coriander
1/4 teaspoon allspice
1/2 cup butter or margarine
1 cup plain yogurt
1/2 cup buttermilk
1 egg, slightly beaten
1 cup currants
1 Tablespoon caraway seeds (optional)

Preheat oven to 350°. Sift the dry ingredients in a large bowl: white and wheat flour, sugar, baking soda, salt, cardamon or coriander, and allspice. Cut in the butter with a pastry fork to make a coarse pastry.

Combine the yogurt, buttermilk, and egg; add to the flour mixture. Fold in the currants and caraway seeds; blend well. Knead until smooth, then divide in half and shape into 2 round loaves about 8 inches in diameter. Flatten and make 2 intersecting slashes across the top. Bake about 1 hour, or until a cake tester comes out clean. Cool. Serve in wedges with butter, cream cheese, or clotted cream.

ISRAELI COFFEECAKE

"Ugatb Kafe"–for morning or mid-afternoon.
Serves 16.

2 cups sugar
1/2 cup butter or margarine
4 eggs
3 cups flour
2 1/2 teaspoons double-acting baking powder
2 teaspoons baking soda
1/4 teaspoon salt
1 cup sour cream
1 cup plain yogurt
2 teaspoons vanilla

Topping:
    1/2 cup sugar
    1 1/2 teaspoons cinnamon
    1 cup walnuts, chopped

Preheat oven to 350°. Cream the sugar and butter together until smooth. Add the eggs one at a time; blend.

Sift the flour with the baking powder, baking soda, and salt. Mix together the sour cream-yogurt. Add the dry ingredients alternately with the sour cream and the yogurt combination to the butter-sugar. Stir in the vanilla. Pour half of the batter into a greased 9 × 13 inch pan.

Prepare the topping by combining the sugar, cinnamon, and walnuts. Put half of the topping on the batter already in the pan. Pour the rest of the batter into the pan and add the rest of the topping.

Bake for about 45 minutes or until the cake tests done.

ITALIAN ZUCCHINI BREAD

This summer squash was first developed in Italy.
Makes 6 to 8 servings.

3 cups flour (can use half whole wheat flour)
1 cup brown sugar
4 teaspoons double-acting baking powder
1 teaspoon baking soda
1 teaspoon ground cinnamon
1 teaspoon salt
1/4 teaspoon allspice
1 1/2 cups zucchini, unpeeled and shredded
1 cup raisins
1 egg, slightly beaten
1 cup plain yogurt
1/4 cup milk
1/4 cup oil
1/2 cup walnuts, chopped

Preheat oven to 350°. Sift the flour with the brown sugar, baking powder, baking soda, cinnamon, salt, and allspice in a large bowl. Stir in the shredded zucchini and raisins. Beat the egg into the yogurt, then add the milk and oil. Fold into the zucchini mixture and add the chopped walnuts.

Pour the batter into a greased 9 × 5 inch loaf pan and bake until a cake tester comes out clean, about 1 hour. Cool in the pan for 10 to 15 minutes, then remove and cool completely on a wire rack.

This bread is perfect with cream cheese and jam. It keeps beautifully, and can be frozen.

KERRY APPLE CAKES

A specialty from my father's Keating heritage in Ireland.
Makes 1 dozen.

2 cups flour
1 cup sugar
2 Tablespoons double-acting baking powder
1 teaspoon salt
1/2 teaspoon baking soda
1/2 teaspoon ground cinnamon
1/4 teaspoon grated nutmeg
1/4 teaspoon allspice
1/4 teaspoon grated lemon rind
1 egg, slightly beaten
1 cup plain yogurt
4 Tablespoons butter or margarine, melted
1 cup apples, pared, cored, and finely chopped

cinnamon-sugar, as garnish

Preheat oven to 400°. Sift the flour with the sugar, baking powder, salt, baking soda, cinnamon, nutmeg, allspice, and grated lemon rind. Put in a large bowl, making a well in the center. Into the well put the egg, yogurt, melted butter, and chopped apples. Stir together just enough to moisten.

Spoon evenly into 12 muffin cups that have been either greased or lined with paper baking cups. Sprinkle with cinnamon-sugar and bake until done, about 15 to 20 minutes. Serve with Irish coffee!

NEW ENGLAND BLUEBERRY MUFFINS

A variation of a recipe from Blueberry Hill Farm, Middlebury, Vermont.
Makes 2 dozen.

2 eggs, slightly beaten
1 cup brown sugar
1 cup plain yogurt
2 Tablespoons butter or margarine, melted
2 cups white flour
1 cup whole wheat or graham flour
2 Tablespoons double-acting baking powder
1/2 teaspoon baking soda
1/2 teaspoon salt
1 pint fresh blueberries
2 Tablespoons flour

Topping:
   sugar
   nutmeg

Preheat oven to 425°. Add the brown sugar and yogurt to the slightly beaten eggs. Stir in the melted butter. Sift the flour with the baking powder, baking soda, and salt. Quickly fold the dry ingredients into the sugar-yogurt mixture, then add the blueberries that have been tossed with 2 Tablespoons flour.

Divide the batter evenly among 24 greased or paper-lined muffin cups. Sprinkle with a little sugar and nutmeg and bake until done, about 12 to 15 minutes.

These muffins can be frozen and reheated quickly in a hot oven.

NEW ZEALAND DOUBLE-SPICE MUFFINS

The delicate, spicy, molasses flavor of these muffins is a native specialty.
Makes 2 dozen.

1 cup butter or margarine
1 cup granulated sugar
4 eggs
1 cup plain yogurt
1 cup dark molasses
4 cups flour
2 teaspoons double-acting baking powder
2 teaspoons baking soda
2 teaspoons ground cinnamon
1 teaspoon grated nutmeg
1/2 teaspoon salt
sugar

Preheat oven to 400°. Cream the butter and sugar with an electric mixer until light and fluffy. Gradually beat in the eggs, one at a time, then add the yogurt and molasses.

Sift the flour with the baking powder, baking soda, cinnamon, nutmeg, and salt. Blend into the batter. Fill 2 dozen paper-lined muffin cups half-full and sprinkle lightly with sugar. Bake about 15 minutes, or until the muffins test done. Remove from the tins and let cool.

NORWEGIAN WALNUT BREAD

They call it "Norsk Nott Brod"—it's super with soups.
Makes 2 loaves.

1 cup white flour, sifted
1 cup whole wheat flour, unsifted
1 teaspoon baking soda
1 teaspoon salt
1/2 cup walnuts, chopped
1/2 cup dates, pitted and chopped
1 egg, slightly beaten
1 cup plain yogurt
2 Tablespoons oil

Preheat oven to 350°. Mix together the sifted white flour, un-sifted whole wheat flour, baking soda, and salt. Add the chopped walnuts and dates; blend. Stir the egg and the oil into the yogurt, then blend into the flour mixture.

Spoon evenly into 2 greased 12-oz. cans and bake for 40 to 45 minutes, or until golden and done. Loosen the edges and turn out onto a wire rack to cool. Serve with butter or cream cheese.

ORANGE-ALMOND CAKE

Spaniards call this orange liqueur specialty "Torta de Almendra."
Makes 16 servings.

1 cup butter or margarine
1 cup sugar
4 eggs, slightly beaten
grated rind of 1 orange (about 1 teaspoon)
1 teaspoon almond extract
1 cup plain yogurt
2 1/2 cups flour
1 teaspoon double-acting baking powder
1 teaspoon baking soda
1/4 teaspoon salt

Syrup:
   1/4 cup sliced almonds
   juice of 1 orange (about 2 Tablespoons)
   1/2 cup sugar
   1/4 cup orange liqueur

Preheat oven to 350°. Cream the butter and sugar together in a bowl, then add the eggs, grated orange rind, almond extract, and yogurt.

Sift the flour with the baking powder, baking soda, and salt; fold into the butter mixture. Pour the batter into a greased tube pan and bake until the cake tests done, about 1 hour.

Let the cake cool on a rack for about 15 minutes, then remove, pierce in several places, and pour the *syrup* topping over the whole cake: mix together the almonds, juice of 1 orange, sugar, and orange liqueur.

Cool completely before serving.

PORTUGUESE CINNAMON CAKE

This popular Iberian cinnamon specialty is known as "Bolo Pardo."
Makes 9 servings.

1/3 cup butter
2 cups granulated sugar
4 eggs
1 cup plain yogurt
3/4 cup milk
4 1/3 cups flour, sifted
1 Tablespoon ground cinnamon
1 teaspoon baking soda

confectioner's sugar, as topping

Preheat oven to 400°. In a large bowl cream the butter and sugar. Add the eggs, one at a time; blend until lemon-colored. Stir in the yogurt and milk.

Sift the flour with the cinnamon and baking soda, then beat into the batter. Turn the mixture into a greased 9-inch tube pan and bake 45 to 60 minutes. Cool for at least 10 minutes, then turn out onto a wire rack. Sprinkle with confectioner's sugar before serving.

PUERTO RICAN SPICECAKE

"Rizcocho de Especies" keeps its Maderia flavor a long time.
Makes 16 servings.

3 1/2 cups cake flour, sifted
1 Tablespoon double-acting baking powder
1 Tablespoon ground cinnamon
1 Tablespoon ground cloves
1 Tablespoon grated nutmeg
1 teaspoon baking soda
1 cup butter or margarine
2 cups dark brown sugar
6 eggs
1 cup plain yogurt

1/3 cup Madeira wine
confectioner's sugar, as topping

Preheat oven to 350°. Sift the flour with the baking powder, cinnamon, cloves, nutmeg, and baking soda; set aside. Cream the butter in a large mixing bowl, gradually adding the brown sugar; beat until well blended. Add the eggs, one at a time, and beat well. Add the flour mixture and the yogurt alternately to the butter mixture, then add the Madeira wine.

Pour into a 9-inch tube pan and bake for 1 hour, or until a cake tester comes out clean. Cool for 10 minutes, then remove from the pan and invert on a wire rack.

When the spicecake is cool, sprinkle with confectioner's sugar.

SCANDINAVIAN OATMEAL PANCAKES

Special "Plattar" pans are available for making these pancakes.
Makes about 1 dozen.

1 egg, beaten
1 cup plain yogurt
1/2 cup quick-cooking oats
1/2 cup flour
2 Tablespoons sugar
2 Tablespoons oil
1 teaspoon double-acting baking powder
1/2 teaspoon baking soda
1/2 teaspoon salt
1/4 teaspoon ground cinnamon
milk (optional)

lingonberry preserves or fruit syrup

In a bowl, stir the yogurt into the egg, then add the oats, flour, sugar, oil, baking powder, baking soda, salt, and cinnamon. (If you want a thinner batter, add some milk.)

Pour about 3 Tablespoons of the batter onto a heated griddle or "plattar" and cook until puffed and lightly golden on both sides.

These nutritious pancakes are traditionally served with lingonberry preserves or fruit syrup.

SCOTTISH SCONES

These teatime treats are served daily throughout the British Isles.
Makes 1 dozen scones.

1/2 cup currants
2 Tablespoons flour
2 1/2 cups flour, sifted
1/2 cup granulated sugar
2 teaspoons double-acting baking powder
1 teaspoon baking soda
1 teaspoon salt
1/4 teaspoon allspice
1/3 cup shortening or lard
1 egg, slightly beaten
1 cup plain yogurt
grated rind of 1 lemon
melted butter

Preheat oven to 425°. Measure out the currants and shake with 2 Tablespoons of the flour to cover; set aside. Sift the remaining flour with the granulated sugar, baking powder, baking soda, salt, and allspice together in a bowl. Cut in the shortening with a pastry fork until the mixture resembles coarse meal. Add the beaten egg, then the yogurt, currants, and lemon rind.

Mix well, then roll out the dough to 1/4-inch thickness. Cut into 2- to 4-inch squares with a pastry wheel, brush with melted butter, then fold each square in half, making a triangle. Pinch the edges together with two fingers. Bake on lightly greased cookie sheets until golden brown, about 12 to 15 minutes. Serve at once, with butter and jam.

SOPA PARAGUAYAN

People in Paraguay spice their cornbread with onions and cheese.
Makes 6 to 8 servings.

1 cup yellow cornmeal
1 cup flour, sifted
1/3 cup granulated sugar
1 Tablespoon double-acting baking powder
1/2 teaspoon baking soda
1/2 teaspoon salt
2 eggs, slightly beaten
1 cup plain yogurt
1 cup butter or margarine, melted
2 teaspoons vegetable oil
1 onion, chopped
1 cup Monterey Jack cheese, grated
4 jalapeño peppers, minced

Preheat the oven to 425°. Blend together the cornmeal, flour, sugar, baking powder, baking soda, and salt in a large bowl. Make a well in the center and pour in the beaten eggs. Add the yogurt and beat with a wooden spoon to blend. Fold in the melted butter.

Brown the onion in the oil until transparent. Add the onions, cheese, and Jalapeño peppers to the cornmeal batter. Pour into a greased 8-inch square baking pan and bake for 25 to 30 minutes, or until the top is lightly browned. Cool for a few minutes, then cut into squares.

SOUTHERN SPOON BREAD

Old-fashioned cornmeal bread is perfect with country ham.
Makes 4 to 6 servings.

1 cup light cream or evaporated milk
1 cup plain yogurt
1/4 cup butter or margarine
2 Tablespoons granulated sugar
1/2 teaspoon salt
1 cup white cornmeal, sifted
4 eggs, separated
1 teaspoon double-acting baking powder
1/8 teaspoon pepper

Preheat oven to 375°. Combine the cream, yogurt, butter, sugar, and salt in a saucepan and heat to the scalding point–but don't boil. Remove from heat and stir in the cornmeal; whip until thick. Beat the egg yolks slightly with the baking powder and pepper. Blend a little of the hot cornmeal mixture with the beaten eggs, then stir the yolks into the mixture in the pan. Beat the egg whites to form soft peaks, then fold gently into the cornmeal.

Pour the batter into a greased, 2-quart casserole and bake, uncovered, until puffed and golden brown. Serve immediately, slathered with lots of butter, salt, and pepper.

SPANISH RUM CAKE

"Bizchoco con Crema" has an incredible creamy filling.
Makes 8 to 10 servings.

5 eggs, separated
1/3 cup sugar
dash of salt
juice and rind of 1 lemon
1/2 cup flour

Filling:
    1 can (11 oz.) condensed cheddar cheese soup
    1 cup plain yogurt
    1/2 cup sugar
    1/3 cup cornstarch
    3 Tablespoons dark rum
    1/2 teaspoon vanilla extract
    3 egg yolks, slightly beaten

confectioner's sugar, as topping

Preheat oven to 400°. Beat the 5 egg yolks together with the sugar, salt, and lemon juice and rind until thick and lemon-colored. Stir in the flour. In a separate bowl, beat the 5 egg whites until stiff, then gently fold into the batter. Pour into 2 greased and floured 8-inch square cake pans and bake for 15 minutes. Remove and cool.

Make the *filling:* in a saucepan blend the soup, yogurt, sugar, cornstarch, rum, and vanilla; cook over low heat until thickened. Pour some of the yogurt mixture into the 3 egg yolks, mix until smooth, and then return to the heat and cook until thickened. Remove, cover with waxed paper, and cool to room temperature.

When ready to assemble, cut each of the cake layers in half horizontally and spread the filling between the layers. Stack and top with confectioner's sugar.

SWAHILI SODA BREAD

An exciting addition from one of my many foreign students.
Makes 9 servings.

1/2 cup butter
1/2 cup granulated sugar
2 eggs
1 cup buttermilk
1 cup plain yogurt
3 cups flour, sifted
3 Tablespoons double-acting baking powder
1/4 teaspoon baking soda
pinch of salt
1 cup raisins

Preheat oven to 350°. Cream the butter and sugar; blend in the eggs. Mix well, then add the buttermilk and yogurt. In a separate bowl, combine the flour, baking powder, baking soda, and salt; add a little at a time to the batter. Add the raisins.

Pour into a greased 9-inch square baking pan and bake for about 1 hour, or until the bread tests done.

SWEDISH PANCAKES

These large baked pancakes are called "Ugnspannkaka."
Makes 6 servings.

2 eggs, beaten
1 1/2 cups milk
1 1/2 cups flour
1 cup plain yogurt
2 teaspoons sugar
1/2 teaspoon salt
1/4 teaspoon ground cinnamon

Topping:
    confectioner's sugar
    fruit purée or jam

Preheat oven to 400°. In a bowl, blend the milk with the eggs, then stir in the flour to make a smooth batter. Add the yogurt, sugar, salt, and cinnamon and blend; let stand for 10 minutes.

Pour the batter into a buttered 9-inch square baking dish and bake for about 20 minutes, or until the pancake is golden brown and puffy. Sprinkle with confectioner's sugar and serve with fruit purée or jam.

SWISS CHERRY CAKE

The Swiss use native cherries for their "Kirschenkuchen."
Serves 10 to 12.

12 ladyfingers, split
2 cups plain yogurt
1 can (14 oz.) sweetened condensed milk
1 teaspoon almond extract
1 can (1 lb., 5 oz.) cherry pie filling

almonds, unblanched and ground
whipped cream

Arrange the ladyfingers standing up around the edge and along
the bottom of a greased 10-inch pie plate. With an electric mixer,
beat the yogurt at high speed, gradually adding the sweetened condensed milk. Beat until bubbly. Fold in the cherry pie filling, spoon
onto the ladyfingers, and freeze until firm.

Take out of the freezer about 20 minutes before serving. Sprinkle
with ground almonds and top with whipped cream.

TRINIDADIAN ORANGE-COCONUT BREAD

Serve this at teatime with Paraguayan maté.
Makes 2 loaves.

3 cups flour
1 Tablespoon double-acting baking powder
1 teaspoon baking soda
1 teaspoon salt
1 cup light brown sugar
2 cups coconut, grated
1 Tablespoon grated orange peel
1/3 cup orange juice
1 egg, slightly beaten
1 cup plain yogurt
1 teaspoon vanilla extract
1/2 cup butter, melted
granulated sugar, as topping

Preheat oven to 350°. Sift the flour with the baking powder, baking soda, and salt in a bowl. Stir in the brown sugar, coconut, orange peel, orange juice, egg, yogurt, vanilla, and melted butter; mix together.

Pour the batter evenly into 2 greased 8 × 4 inch loaf pans, sprinkle with sugar, and bake until the bread tests done, about 1 hour. Cool for a few minutes in the pans, then turn out onto wire racks.

Lunches, Suppers, Dinners

A man once asked Diogenes what was the proper time for supper, and he made the answer, "If you are a rich man, whenever you please; and if you are a poor man, whenever you can."

Whenever you decide to have these entrees, you're guaranteed to exclaim how easily exquisite they are when made with yogurt. Whether the yogurt is used in tenderizing marinades, as flavor enhancers, or in sauces and gravies, it is uniquely satisfying.

Chicken is particularly suited to yogurt. *African Chicken-Rice Stew* (Jollof Rice) has rice cooked right in the casserole, while *Arroz con Pollo,* a favorite of Spanish-speaking people all over the world, adds ham and special seasonings to the chicken-rice combination. *Raved-Over Romanian Chicken* will become one of your family favorites, vying with *Balinese Braised Chicken* for first prize.

Hungarians serve their *Chicken Paprikash* with either noodles or dumplings. "Tandoori Murghi," *Indian Roast Chicken,* is a popular dish in India and Pakistan. *Malay Coconut Chicken* (Rendan Santan) uses distinctive Far Eastern spices. Make lots of *Moroccan-Style Chicken;* it's a real crowd-pleaser. Eggplant is the secret to *Yugoslavian Chicken and Asparagus.* And it should be noted that *Thai Chicken Curry* is a favorite of our Siamese cats, who also like leftovers of *Antilles Chicken Livers.*

Beef is called *Bif Stroganov,* a nineteenth-century Russian creation honoring Count Paul Stroganov, when it is served with a mustard-yogurt-mushroom sauce over noodles. From Germany comes spectacular *Sauerbraten. Bobotie,* an African Meat Pie, uses ground beef, as does *Hamburg-Cheese Casserole* and *Mexican Meal-in-a-Minute.* Beef ground together with pork and veal turns an old American standby into a tasty new adventure in *Meatloaf*

139

Italian-Style. Beef chuck as well as beef liver come skewered on *West African Beef-Liver Kabobs*, and the Hungarian steaks called *Eszterhazy Rostelyos* are often requested at my house.

Eggplant combines well with beef in *Moussaka Yugoslavian-Style*, while zucchini takes the honors in Greek *Pastitsio Makaronia Me Feta*. Lots of (leftover) vegetables are featured in my prize-winning *Yankee Red Flannel Hash*.

Pork, in the form of frankfurters (hotdogs), is a tradition in my home state of Massachusetts on Saturday nights; that variation here is named *Boston-Baked Lima Beans 'n' Franks*. Or, you might try *Dutch Potatoes and Bratwurst*. The British have great fun with sausages in *English Toad-in-the-Hole*. Italians use their sausages whole or cut up in *Italian Heroes*, which are especially popular with the teenage crowd.

Another pork dish that's ideal for lots of people is *Quiche Lorraine for a Crowd*. "Veprove s Krenem," *Stuffed Czech Chops*, are Czechoslovakia's answer to pork with horseradish, served with caraway noodles. The *Yam-Ham Pie* from the Dominican Republic is always a treat.

Ground pork and veal team up in *Canadian Christmas Pie*, veal being another natural with yogurt. The Swiss have given us *Emince de Veau*, which is a must for your table. Another variation, "Emince de Veau avec Yaourt," features pasta: *Swiss Spaghetti-Veal Casserole*.

Lamb that is ground, mixed with spices, and put into pita bread is a Greek contribution called *Gyros*. *Pakistani Leg of Lamb* is served with a cucumber sauce, while a variation for the Greek preference for a spicy yogurt sauce (Arni Me Yaourti) is also included. Serve *Rogan Jaush*, an Indian Epicurean dish meaning "Color-Passion Curry," with hot steamed rice.

Seafood is called "Fruits de Mer" in French; elegant *Creamed Seafood à la Monaco* is served in patty shells, *Chesapeake Bay Scalloped Oysters* and French *Coquilles en Casserole* in ramekins. *Fabulous Philippine Fish* (Escabeche) uses red snapper, bluefish, or bass.

Stuffed sole (*French Filets de Sole Farcis à la Creme*) is a year-round delight for an intimate dinner party of six; the shrimp, wine and cognac, and grated Swiss cheese blend beautifully with yogurt.

Alaskan Crab Casserole is nice for lunch or dinner, as is *Madras Shrimp.*

Polish Pike features a horseradish sauce that can be used on any number of fish dishes, while *Icelandic Atlantic Fish* favors a yogurt-mustard topping. *Hong Kong Halibut* is always a nice spicy change for white fish. And if you can get ahold of salmon, you'll love *Le Soufflé Saumon Canadien* (Canadian Salmon Soufflé).

As you can see, yogurt makes your entrees–whether for lunches, suppers, or dinner–delightfully delicious.

LUNCHES, SUPPERS, DINNERS

AFRICAN CHICKEN-RICE STEW

"Jollof Rice" means the rice is cooked in this colorful, spicy casserole.
Serves 4.

2 Tablespoons peanut oil
2 onions, chopped
1 green pepper, seeded and sliced into rings
1 teaspoon curry powder
1 teaspoon salt
1 teaspoon sugar
1/2 teaspoon garlic powder
1/2 teaspoon ground ginger
2 whole cloves
dash of hot pepper sauce
1 can (15 oz.) stewed tomatoes, chopped
1/2 lb. fresh green beans
1/2 cup uncooked rice
1 cup plain yogurt
2 cups chicken, cooked and cubed
1 teaspoon lemon juice

Sauté the onions and green pepper in the oil until softened, about 5 minutes. Stir in the curry powder, salt, sugar, garlic powder, ginger, cloves, and hot pepper sauce. Add the stewed tomatoes, green beans, and rice; cook 15 minutes.

Blend in the yogurt, cooked chicken, and lemon juice; reheat and serve.

ALASKAN CRAB CASSEROLE

Crabmeat from Alaska blends beautifully with yogurt.
Serves 4.

1 1/2 cups packaged stuffing mix
1 cup plain yogurt
6 hard-cooked eggs, peeled and mashed
12 oz. crabmeat, fresh or canned
1 cup mayonnaise
2 Tablespoons fresh parsley, chopped
1/4 cup onion, minced (optional)

Preheat oven to 350°. Soften 1 cup of the stuffing mix with the yogurt. Add the hard-cooked eggs, crabmeat, mayonnaise, parsley, and onion.

Place this mixture into a greased 1 1/2-quart casserole, then top with the remaining 1/2 cup stuffing, and bake for 30 minutes.

ANTILLES CHICKEN LIVERS

Nutrition-conscious eaters will appreciate this West Indian flavor.
Serves 4.

4 slices bacon, cooked
2 onions, chopped
1 clove garlic, minced
1 green pepper, seeded and cut into strips
1 lb. chicken livers, halved
1 can (10 1/2 oz.) cream of mushroom soup
1 cup plain yogurt
2 teaspoons soy sauce
1/2 teaspoon salt
1/4 teaspoon pepper
2 hard-cooked eggs, sliced
1 cup cooked peas (optional)

Cook the bacon; remove and crumble. Sauté the onions, garlic, and green pepper in a small amount of the bacon grease, draining off the rest. Add the chicken livers and cook until tender, about 5 to 10 minutes, stirring occasionally. Turn heat to low and add the soup, yogurt, soy sauce, salt, and pepper. Slowly heat through.

Add the sliced eggs and peas, then pile the mixture on toast or rice. Garnish with the bacon bits.

ARROZ CON POLLO

Spanish-speaking people serve rice-chicken-ham variations.
Serves 6.

3 to 4 lbs. chicken, cooked and cut-up
1 cup ham, diced
2 cups rice, cooked
1/2 teaspoon ground coriander
1/2 teaspoon salt
1/4 teaspoon pepper
1 cup plain yogurt
1/2 cup sour cream
1 cup tomatoes, peeled, seeded, and chopped
1 green pepper, seeded and chopped
1 Tablespoon capers
1/4 cup green olives, pitted and sliced
8 oz. Monterey Jack cheese, grated

Preheat oven to 350°. Mix together the chicken, ham, rice, coriander, salt, and pepper. Put about one-third of the mixture into a greased 1-quart casserole.

Stir the yogurt and sour cream together; add the tomatoes, green pepper, capers, and olives. Put half of this mixture in the casserole, top with chicken-ham-rice, then grated cheese; repeat this process. Bake until the cheese on top has melted, about 30 minutes.

BALINESE BRAISED CHICKEN

Our son Keith was an exchange student on the beautiful island of Bali.
Serves 4.

1 frying chicken, 2 to 3 lbs., cut up
2 Tablespoons peanut oil
1 onion, thinly sliced
1 clove garlic, minced
2 teaspoons ground coriander
1 teaspoons ground turmeric
1/2 teaspoon chili powder
1/4 teaspoon grated ginger
1/4 teaspoon salt
dash of black pepper
1 Tablespoon flour
1 Tablespoon lemon or lime juice, freshly squeezed
1 cup plain yogurt

rice

Heat the peanut oil in a frying pan and brown the chicken pieces on both sides; remove. Add the onion and garlic and sauté until tender, then stir in the coriander, turmeric, chili powder, ginger, salt, and pepper. Cook 1 minute.

Return the chicken to the pan. Combine the flour, lemon or lime juice, and yogurt; add slowly to the chicken. Cover the skillet, then cook slowly until done, about 30 to 40 minutes. Serve over hot, cooked rice.

BIF STROGANOV

A nineteenth-century Russian creation honoring Count Paul Stro-
ganov.
Serves 4.

1 1/2 lbs. top sirloin of beef
1 teaspoon salt
1/2 teaspoon pepper
2 onions, chopped
2 Tablespoons butter or margarine
1 Tablespoon flour
1 cup beef stock
1 teaspoon prepared mustard (preferably hot)
1 cup plain yogurt
1/2 lb. mushrooms, sliced

parsley
noodles

Remove any fat from the sirloin and cut it into narrow strips, 2 inches long and 1/2 inch thick. Sprinkle with salt and pepper, then brown quickly with the onions in the butter; remove the meat and onions.

Blend the flour into the gravy, then add the beef stock; cook until thickened. Stir in the mustard, yogurt, and mushrooms and heat slowly. Sprinkle with parsley and serve with noodles.

BOBOTIE (African Meat Pie)

This meat pie is a special regional favorite.
Serves 6 to 8.

2 slices of bread (preferably wheat)
1 cup plain yogurt
2 onions, chopped
2 Tablespoons oil
2 lbs. beef, finely ground
2 Tablespoons plum jam
1 bay leaf
1 teaspoon salt
1 teaspoon curry
2 eggs

Garnishes:
   lemon slices
   pimento strips

Preheat oven to 350°. Soak the bread in yogurt for a few minutes, then squeeze out the yogurt and mash up the bread. Save the leftover yogurt. Sauté the onions in the oil and mix in the bread. Add the ground beef, plum jam, the bay leaf, salt, curry, and 1 egg. Shape into a round heap and put in a 9-inch pie plate.

Beat the second egg into the leftover yogurt, add another pinch of salt, and pour over the meat. Bake until cooked through, about 1 hour. Drain off any excess liquid, then top with lemon slices and pimento strips.

BOHEMIAN VEAL ROAST

Spaetzle dumplings go especially well with this rich gravy.
Serves 8.

5 lbs. veal rump roast, boneless
2 Tablespoons oil
1 cup plain yogurt
1/2 cup beef stock
2 onions, chopped
2 teaspoons fresh dill seed
1 teaspoon salt
1/2 teaspoon pepper

Brown the veal roast on all sides in the oil in a heavy kettle. Combine the yogurt, beef stock, onions, dill seed, salt, and pepper; spread over the top and sides of the roast. Then you have several choices for cooking:

1. Cover and cook in the same kettle over low heat until the meat is tender, about 3 hours.
2. Put in a slow-cooking pot, allowing 8 to 10 hours for the roast to be cooked through.
3. Bake in a 325° oven until tender, about 2 hours.

When the veal roast is done, arrange it on a hot platter, cover with pan gravy, then carve and serve.

BOSTON-BAKED LIMA BEANS 'N' FRANKS

A variation on the traditional Bay State Saturday night supper.
Servse 8 to 10.

1 lb. dried lima beans
1 teaspoon salt
1/2 cup brown sugar
2 Tablespoons molasses
1 Tablespoon dry mustard
2 teaspoons salt
1 cup plain yogurt
1/2 cup butter or margarine

1 lb. frankfurters, boiled or fried

Soak the lima beans overnight in water to cover. The next day, add the salt and cook in the soaking water until tender, about 1 hour. Drain and rinse, then put in a bean pot with the brown sugar, molasses, dry mustard, salt, yogurt, and butter. Bake in a 300° oven for about 2 hours, uncovering the pot the last hour. This can be made ahead and reheated. Serve with the cooked franks.

Note: Greek lima beans with yogurt (Koukia Me Yaourti) are cooked with 1 chopped shallot, 1/4 cup parsley, 2 Tablespoons dill or mint leaves, *1 cup plain yogurt*, 1/2 teaspoon sugar, and salt and pepper to taste. Bake as above.

BULGARIAN BAKED HASH

"Musaka" is popular at many Black Sea homes and resorts.
Serves 6 to 8.

2 lbs. ground beef
2 onions, chopped
1 Tablespoon fresh parsley, chopped
1 teaspoon salt
1/4 teaspoon pepper
dash of cayenne pepper
1 8-oz. can tomato sauce
8 medium-sized potatoes, peeled and diced

1 cup plain yogurt
2 eggs, slightly beaten
2 Tablespoons flour

Preheat oven to 375°. Sprinkle a large frying pan with salt and brown the ground beef and onion; add the parsley, salt, pepper, cayenne pepper, and tomato sauce. Simmer for about 15 minutes, stirring occasionally. Add the potatoes and simmer another 10 minutes.

Turn the beef-potato mixture into a shallow casserole dish and bake until browned, about 30 minutes. Stir the eggs and flour into the yogurt and pour over the browned hash. Return to the oven for about 10 more minutes.

CANADIAN CHRISTMAS PIE

"Tourtiere de Noel" is traditionally served on Christmas Eve.
Makes 6 to 8 servings.

2 9-inch prepared pie crusts
2 Tablespoons vegetable oil
2 onions, chopped
1 clove garlic, minced
2 lbs. ground veal and pork, mixed
1 Tablespoon flour
2 teaspoons Worcestershire sauce
1 teaspoon garlic salt
1/4 teaspoon pepper
1/4 teaspoon allspice
1 cup plain yogurt
2 cups chicken or turkey, cooked and cubed
1 egg, beaten

Preheat oven to 375°. Sauté the onions and garlic in the oil until tender, then add the mixed ground veal and pork and cook until browned. Blend in the flour, Worcestershire sauce, garlic salt, pepper, allspice, and yogurt. Pour this mixture into one prepared pie crust shell, then cover with the cubed poultry.

Put the other pie crust on top, brush with the beaten egg, and cut some slits in the top to allow steam to escape. Bake for 1 hour, or until the top pastry is golden brown. Let stand a few minutes, then serve with a salad, wishing all a Joyeux Noel.

CHESAPEAKE BAY SCALLOPED OYSTERS

Oysters from the Maryland shore are unparalleled.
Serves 4.

1 cup cheddar cheese, grated
1/4 cup Parmesan cheese, grated
1 cup cornbread, finely crumbled
4 Tablespoons butter, melted
1 teaspoon salt
1/4 teaspoon pepper
1/4 teaspoon ground mace
1 lb. Chesapeake Bay oysters, shelled
1 cup plain yogurt
2 teaspoons sherry (optional)

Preheat oven to 350°. Mix together the cheddar and Parmesan cheeses, cornbread crumbs, butter, salt, pepper, and mace. Put half the mixture into 4 slightly buttered ramekins.

Combine the oysters, yogurt, and sherry and place on top of the cheese-breadcrumb mixture. Cover with the remaining breadcrumb mixture, and bake until the crumbs are browned, about 20 to 25 minutes.

CHICKEN PAPRIKASH

Hungarians serve this easy-to-prepare dish with either noodles or dumplings.
Serves 4 to 6.

3 to 4 lb. chicken, skinned and cut up
2 onions, chopped
2 Tablespoons vegetable oil
1 Tablespoon Hungarian paprika
1 teaspoon salt
1/4 teaspoon pepper
1 cup chicken broth
1 cup plain yogurt

2 Tablespoons fresh parsley, chopped
noodles or dumplings

Sauté the chicken and onions in oil until browned on both sides, about 10 minutes. Stir in the paprika, salt, pepper, and chicken broth. Bring to a boil, cover and simmer until the chicken is tender, about 40 minutes.

Just before serving, stir in the yogurt and sprinkle with parsley. Serve with noodles or dumplings.

COQUILLES EN CASSEROLE

French-style scallops–the most delicate of all shellfish.
Serves 4 as an entree, 6 as an appetizer.

2 lbs. bay or sea scallops
1 cup dry white wine
2 Tablespoons lemon juice
1 can (4 oz.) mushroom caps, drained
2 Tablespoons butter
2 Tablespoons flour
1/2 teaspoon garlic salt
1/4 teaspoon pepper
1/4 teaspoon fresh thyme
1/2 cup Swiss cheese, shredded
1 cup plain yogurt

1/4 cup Parmesan cheese, grated
1/2 cup breadcrumbs

Poach the scallops in a combination of the wine, lemon juice, and mushroom caps for 5 to 10 minutes. Reserve the liquor.

In a separate pot, make a white sauce by melting the butter, stirring in the flour, garlic salt, pepper, and thyme, and then adding the scallop liquor; heat until thickened. Stir in the Swiss cheese and yogurt, then add the scallops. Put into individual lightly buttered ramekins or a shallow, buttered pan, sprinkle with Parmesan cheese and breadcrumbs, and broil until browned.

CREAMED SEAFOOD À LA MONACO

"Bouchees aux Fruits de Mer"–elegant in patty shells.
Serves 6 for luncheon.

6 individual patty shells, baked
2 Tablespoons butter
1 onion, chopped
1 clove garlic, minced
1/2 pound button mushrooms
2 Tablespoons flour
1 cup dry white wine
1 cup plain yogurt
1 teaspoon salt
1/4 teaspoon pepper
1 Tablespoon cognac
1 lb. mussels, cooked and shelled
1/2 lb. shrimp, shelled, deveined, and cooked
1/2 lb. lobster meat, cooked

Prepare the patty shells while making the seafood mixture. In a frying pan, sauté the onions, garlic, and mushrooms in the butter until golden. Stir in the flour and wine and cook until thickened, stirring constantly; add the yogurt, salt, pepper, and cognac.

Next, stir in the mussels, shrimp, and lobster meat. Heat gently, then scoop into the warm patty shells and serve with a tossed salad.

DUTCH POTATOES AND BRATWURST

Edam cheese makes the delightful difference in this entree.
Serves 6 to 8.

6 to 8 potatoes, peeled and thinly sliced
1/2 cup butter
1/2 lb. Edam cheese, grated
2 onions, chopped
1/4 cup fresh parsley, chopped
2 teaspoons salt
1/2 teaspoon pepper
1/2 teaspoon paprika
2 cups plain yogurt
1/4 cup Edam cheese, grated

1 dozen bratwursts

Preheat oven to 450° to start. Put one-quarter of the potatoes in a shallow, buttered, oblong casserole. Dot with 2 Tablespoons butter, then sprinkle with a portion of the following: mix together the 1/2 lb. of grated cheese, onions, parsley, salt, pepper, and paprika. Repeat, making 4 layers.

Bake for 10 minutes, then add the yogurt and sprinkle with 1/4 cup more cheese, reduce oven to 350°, and continue baking for 2 more hours, or until the potatoes are tender.

Serve with the bratwursts, cooked separately. Dutch beer makes a nice accompaniment.

EMINCE DE VEAU

Switzerland's classic creation with veal.
Serves 4 to 6.

2 lbs. veal cutlets
1/3 cup butter
2 Tablespoons fresh shallots, minced
1 Tablespoon flour
1 cup dry white wine
1 cup plain yogurt
1 teaspoon salt
1/4 teaspoon pepper

fresh parsley, for garnish

Trim the veal and cut into thin slivers about 2 inches long. Sauté quickly in the butter until tender, about 2 minutes; remove.

Add the shallots and flour to the pan drippings. Stir until blended, then add the white wine. Bring to a boil, then simmer until the liquid is reduced. Turn down the heat, stir in the yogurt, salt, and pepper, and heat through. Garnish with parsley.

ENGLISH TOAD-IN-THE-HOLE

Have the kids guess where this popular dish got its name.
Serves 4 to 6.

1 dozen pork sausages
2 teaspoons oil
2 eggs, well beaten
1 cup milk
1 cup plain yogurt
1 cup flour
1/2 teaspoon salt
1/4 teaspoon pepper
1/4 teaspoon dry mustard

Preheat oven to 425°. In a 9-inch square pan, cook the sausages in the oil until browned. Drain off the excess fat, but leave the sausages in the pan.

In a bowl, beat together the eggs, milk, and yogurt. Add to a mixture of the flour, salt, pepper, and dry mustard sifted together. Pour this batter over the sausages and bake, uncovered, about 30 minutes. When done, the dish should be puffed and golden brown, resembling a popover with a "toad-in-the-hole."

ESZTERHAZY ROSTELYOS

Hungarian braised sirloin is a classic; a favorite of most men.
Serves 6.

6 sirloin steaks, about 1/2 lb. each, cut 1/2 inch thick
1 teaspoon salt
1/2 teaspoon pepper
2 Tablespoons butter or margarine
6 onions, chopped
6 carrots, peeled and sliced
8 to 10 celery stalks, sliced
2 Tablespoons flour
6 whole peppercorns
1 Tablespoon Hungarian paprika
1 can condensed beef bouillon, undiluted
1 Tablespoon capers
1 lemon, thinly sliced
1 cup plain yogurt

fresh parsley, chopped
egg noodles, cooked

Pound the salt and pepper into the steaks, then brown them on both sides in the butter. Remove from the skillet and set aside. Add the onions, carrots, and celery and sauté about 15 minutes, stirring occasionally. Stir in the flour, then the peppercorns, paprika, and bouillon.

Return the steaks to the skillet, cover, and simmer 30 minutes, or until the vegetables and meat are tender. Add the capers and lemon slices. Cook, uncovered, until the liquid is reduced by 1/3–about 15 minutes. Slowly stir in the yogurt and heat through. Sprinkle with the parsley and serve with egg noodles.

FABULOUS PHILIPPINE FISH

"Escabeche"–you can use red snapper, bluefish, or bass.
Serves 4.

2 lbs. prepared fish–red snapper, bluefish, or bass (heads removed)
1/4 lb. butter, melted
1 cup plain yogurt
1 onion, chopped
1 clove garlic, minced
1 Tablespoon brown sugar
1 Tablespoon soy sauce
1/2 teaspoon celery salt
1/4 teaspoon ground ginger

Preheat oven to 375°. Melt the butter in a pan, then remove from the heat and add the yogurt, onion, garlic, brown sugar, soy sauce, celery salt, and ginger. Spread on both sides of the prepared fish, then put the fish in a shallow, buttered dish and bake until it flakes with a fork, about 20 to 30 minutes.

Serve any extra sauce separately; leftovers can be frozen.

FRENCH FILETS DE SOLE FARCIS À LA CREME

Fantastic French stuffed fillets of sole.
Serves 6.

2 lbs. fillets of sole (about 6)
1/2 teaspoons fine herbes (a French spice specialty)
2 Tablespoons butter
2 onions, chopped
1 clove garlic, minced
1/2 lb. button mushrooms
1/2 lb. shrimp, shelled, deveined, and cooked
2 Tablespoons fresh parsley, chopped
1 cup plain yogurt
1 cup dry white wine
2 Tablespoons cognac
1 teaspoon salt
1/2 cup Swiss cheese, grated

Preheat oven to 400°. Sprinkle the fillets with fine herbes and set aside. Sauté the onions, garlic, and mushrooms in the butter until golden; stir in the shrimp and parsley. Put rounded teaspoons of this mixture into the 6 fillets, roll up, then arrange them into a buttered casserole dish.

Combine the yogurt, wine, cognac, and salt; pour over the fillets. Sprinkle the top with the grated cheese. Bake for 20 to 25 minutes, or until the top is golden brown and the fillets flake easily. All you need is some French bread and a salad.

GYROS

Greek spiced lamb in pita bread is an international favorite.
Serves 6 to 8.

2 lbs. ground lamb
2 onions, chopped
1 clove garlic, minced
1 teaspoon allspice
1 teaspoon ground coriander
1 teaspoon dried savory
1/2 teaspoon salt
1/4 teaspoon pepper

pita bread

Garnishes:
    2 tomatoes, peeled and chopped
    1 cup plain yogurt
    1 cup fresh parsley, chopped
    oil and vinegar

Brown the lamb, onions, and garlic together; drain off any excess fat. Add the allspice, coriander, savory, salt, and pepper.

Pile the lamb mixture into pita bread and garnish with the tomatoes, yogurt, parsley, and oil and vinegar to taste.

HAMBURG-CHEESE CASSEROLE

Our family refers to this favorite supper simply as "The Casserole."
Serves 6 to 8.

salt
2 lbs. ground beef
1 can (15 1/2 oz.) tomato sauce
1 cup small curd cottage cheese
8 oz. cream cheese, softened
1 cup plain yogurt
1 onion, chopped
1 green pepper, seeded and chopped
8 oz. package egg noodles, cooked

Preheat oven to 350°. Sprinkle some salt on the bottom of a large frying pan; add the ground beef, and brown at medium heat. Stir in the tomato sauce; set aside.

Mix together the cream cheese, cottage cheese, and yogurt. Add the chopped onion and green pepper.

Put half the cooked egg noodles into a greased 2-quart casserole. Cover with the cheese mixture, then add the rest of the noodles. Top with the hamburg-tomato combination. Bake for 25 to 30 minutes or until heated through.

Add some Italian bread and a green salad, and the gang will rave.

HONG KONG HALIBUT

This is a spicy fish dish from the culinary capital of the world.
Serves 4.

1 lb. halibut (or other white-fleshed fish)

Marinade:
  1 cup plain yogurt
  2 teaspoons ground ginger
  1/2 teaspoon chili powder
  1/2 teaspoon turmeric

  2 Tablespoons butter or margarine
  1 clove garlic, minced
  1 onion, chopped
  1 teaspoon sugar
  1/2 teaspoon ground cinnamon
  1/2 teaspoon ground cloves
  1/2 teaspoon ground cardamon

Divide the fish into 4 equal portions, place in a shallow dish, and cover with a mixture of the yogurt, ginger, chili powder, and turmeric. Cover with plastic wrap and marinate in the refrigerator for about an hour.

Melt the butter in a frying pan, then sauté the garlic and onion in it until tender. Stir in the sugar and spices (cinnamon, cloves, and cardamon) and cook 1 minute. Add the fish and yogurt marinade mixture and cook slowly over medium heat until the fish is tender, about 5 to 10 minutes.

ICELANDIC ATLANTIC FISH

Fish is the #1 resource for these Norwegian Viking descendants.
Serves 4.

2 lbs. fish fillets (such as cod, haddock, sole, and/or halibut)
juice of 1 lemon
1/2 teaspoon salt
1/4 teaspoon pepper
1/4 teaspoon paprika
8 oz. Icelandic cheese, grated (available at most cheese stores)
2 teaspoons dry mustard
1 cup plain yogurt
1/2 cup breadcrumbs, preferably from dark bread

Preheat oven to 350°. Sprinkle the fish on both sides with a combination of the lemon juice, salt, pepper, and paprika; put into a shallow, buttered baking dish. Top with the grated cheese. Stir the dry mustard into the yogurt, then pour the yogurt-mustard mixture on top of the fish. Top with the breadcrumbs.

Bake for 30 to 35 minutes, or until the fish flakes easily with a fork.

INDIAN ROAST CHICKEN

"Tandoori Murghi" is a popular yogurt-marinated dish in India and Pakistan.
Serves 6 to 8.

4 lb. broiler-fryer chicken, cut up

Marinade:
   4 Tablespoons lemon juice
   1 teaspoon prepared mustard
   1 teaspoon garlic salt
   1 cup plain yogurt
   1/2 teaspoon ground cardamon
   1/4 teaspoon ground ginger
   1/4 teaspoon ground cumin
   1/4 teaspoon black pepper
   dash of red pepper sauce

Preheat oven to 375°. Arrange the chicken portions in a large bowl and top with the *marinade*: mix together the lemon juice, prepared mustard, garlic salt, yogurt, cardamon, ginger, cumin, pepper, and red pepper sauce. Coat well, cover, and refrigerate for 12 to 24 hours.

When ready to cook, remove the chicken pieces and put them into a shallow roasting pan. Bake, uncovered, basting occasionally with the marinade, for about 2 hours.

ITALIAN HEROES

These Italian sausages are popular with the teenage crowd.
Makes 6.

6 Italian sausages, mild or hot

Sauce:
    2 Tablespoons olive oil
    2 onions, chopped
    2 green peppers, seeded and chopped
    1 clove garlic, minced
    3 oz. can tomato paste
    1 teaspoon dried basil
    1/2 teaspoon dried oregano
    1/2 teaspoon fresh parsley, chopped
    6 hero-style rolls

Toppings:
    1 cup plain yogurt
    shredded lettuce
    chopped tomatoes
    sliced carrots
    Parmesan cheese

Cook the sausages by boiling, frying, or steaming them; put them into the individual hero-style rolls, either whole or sliced.

In a frying pan, make a sauce by browning the onions, green pepper, and garlic in the olive oil. Add the tomato paste, basil, oregano, and parsley. Pour this mixture over the heroes, and top with the yogurt, lettuce, tomatoes, carrots, and/or Parmesan cheese.

MADRAS SHRIMP

Mint chutney makes a nice accompaniment to this shrimp spectacular.
Serves 4.

1 can (10 1/2 oz.) cream of shrimp soup
1 lb. shrimp, shelled, deveined, and cooked
1 cup plain yogurt
8 oz. package egg noodles
1 onion, chopped
1 teaspoon ground turmeric
1/2 teaspoon dried coriander
1/2 teaspoon ground ginger
1/4 teaspoon ground cumin
10 oz. package frozen green beans or snow peas, cooked
1 cup cheddar cheese, grated

Preheat oven to 350°. Mix together the soup, shrimp, and yogurt. Cook the egg noodles as directed, then drain and combine with the shrimp sauce, onion, and spices.

Lightly grease a 2-quart casserole, then add layers of green beans or snow peas and the shrimp-noodle mixture. Top with the grated cheese and bake until bubbly, about 30 minutes.

MALAY COCONUT CHICKEN

*Known as "Rendan Santan," this marinated meal uses distinctive
Far Eastern spices.*
Serves 6 to 8.

2 3-lb. chickens, cut-up

Marinade:
    1 clove garlic, minced
    1 teaspoon dried coriander
    1 teaspoon chili powder
    1 teaspoon ground cumin
    1/2 teaspoon salt
    1/4 teaspoon pepper
    1/4 teaspoon powdered saffron
    1 cup plain yogurt
    1/2 cup fresh coconut, grated

1 lemon, thinly sliced, for garnish

Preheat oven to 350°. Stir the marinade ingredients together: the
garlic, coriander, chili powder, cumin, salt, pepper, saffron, yogurt,
and coconut. Pour over the chicken pieces and marinate in the
refrigerator for at least 1 hour.

Top the chicken with the lemon slices and bake until the chicken
is done, about 40 to 45 minutes.

MEATLOAF ITALIAN-STYLE

The old American stand-by gets a zesty new taste from mozzarella.
Serves 8.

2 lbs. beef, pork, and veal, ground together
1/2 cup wheat germ
1 1/2 cups whole wheat bread crumbs
2 eggs, slightly beaten
1 onion, chopped
1/4 cup catsup
1 Tablespoon Worcestershire sauce
1 teaspoon dried oregano
1 teaspoon fresh basil
2 Tablespoons fresh parsley, chopped
1 teaspoon salt
1/4 teaspoon pepper
1 cup plain yogurt
1/4 lb. mozzarella cheese, cut into small cubes

Preheat oven to 350°. Combine the ingredients in the following order: mixed ground beef, pork, and veal, wheat germ, bread crumbs, eggs, onion, catsup, Worcestershire sauce, oregano, basil, parsley, salt, pepper, and yogurt. Punch the cheese cubes into the mixture.

Bake in a 9 × 5 inch well-packed loaf pan until the meat is cooked, about 1 hour. Top with more catsup, if desired, before turning out, slicing, and serving.

MEXICAN MEAL-IN-A-MINUTE

Be sure to keep your shelves stocked with Mexican mixings.
Serves 4 to 6.

salt
1 lb. lean hamburger meat
2 onions, chopped
1 can (15 oz.) pinto beans, drained
1 can (10 oz.) enchilada sauce
1 can (8 oz.) tomato sauce
1/8 teaspoon chili powder
1 cup Monterey Jack cheese, grated
1 package (6 oz.) corn chips, crushed and divided
1 cup plain yogurt
1/4 cup more Monterey Jack cheese

Preheat oven to 375°. Sprinkle a frying pan lightly with salt, then brown the hamburger meat and onions in it. Drain, remove from the heat, and add the pinto beans, enchilada sauce, tomato sauce, chili powder, cheese, and all but 1 cup of the corn chips.

Put the mixture into a 2-quart casserole and bake for about 30 minutes. Spread the top with yogurt, sprinkle with the remaining Monterey Jack cheese, and decorate with more corn chips. Broil briefly until the cheese melts and the top is lightly browned.

MOROCCAN-STYLE CHICKEN

Crowds are crazy about this spicy supper that can be made in advance.
Serves 6.

1/4 cup olive oil
2 onions, chopped
2 cloves garlic, minced
4 Tablespoons fresh parsley, chopped
2 teaspoons ground ginger
2 teaspoons dried coriander
1/2 teaspoon dried cumin
1/2 teaspoon salt
dash of pepper
2 frying chickens, cut up
1 cup chicken broth
1 lemon, thinly sliced
1 cup green olives, pitted
1 cup plain yogurt

Cook the onions and garlic in the olive oil in a frying pan until tender, then mix in the spices. Fry the chicken pieces until golden brown, then add the chicken broth; cover and cook slowly for 30 minutes. Mix in the lemon slices and olives and cook 20 minutes more.

Just before serving, stir in the yogurt–it makes a nice gravy. This recipe can easily be doubled. Add some rice, pita bread, and a Middle Eastern salad.

MOUSSAKA YUGOSLAVIAN-STYLE

Eggplant and meat combinations are well worth the effort.
Serves 8 to 10.

Meat layer:
    2 Tablespoons vegetable oil
    2 onions, chopped
    1 clove garlic, minced
    2 lbs. beef, pork, and lamb, ground together
    2 eggs
    1 teaspoon salt
    1/4 teaspoon pepper
    1/4 teaspoon allspice
    1 cup dark bread crumbs (preferably rye or pumpernickel)
    1 can (8 oz.) tomato sauce

Eggplant layer:
    2 large eggplants, peeled and cut into 1/4 inch slices
    1/2 teaspoon salt
    1/4 cup flour
    4 eggs, beaten
    oil, for frying

Yogurt crust:
    4 Tablespoons butter
    2 Tablespoons cornstarch
    1 cup milk
    1/4 teaspoon salt
    1/8 teaspoon grated nutmeg
    1 cup plain yogurt
    2 eggs, slightly beaten

Preheat oven to 350°. To make the meat layer: fry the onions and garlic in oil until tender, then add the ground beef, pork, and lamb; cook until browned. Add the eggs, salt, pepper, allspice, breadcrumbs, and tomato sauce. Set aside.

Sprinkle the sliced eggplant with salt, then let it drain about 15 minutes. Sprinkle with flour, dip in the egg, then brown on both sides in the oil. Alternate layers of cooked eggplant and meat mixture in a greased, 4-quart casserole, beginning and ending with the eggplant.

For the yogurt topping, melt the butter over low heat, then add the milk and cornstarch. Cook until thickened, stirring constantly, about 10 minutes. Stir in the salt, nutmeg, and yogurt. Beat the eggs, then stir slowly into the yogurt mixture. Pour over the eggplant-meat combination and bake, uncovered, about 1 hour.

ORIENTAL PORK AND NOODLES

Sweet-and-sour pork with oriental noodles is both an Asian and American choice.
Serves 4.

2 cups pork, cooked and cut into cubes
2 Tablespoons vegetable oil
1 onion, chopped
1 green pepper, seeded and cut into thin strips
1 can (1 lb., 4 oz.) pineapple chunks, drained (reserving juice)
2 Tablespoons brown sugar
2 Tablespoons vinegar
1 Tablespoon cornstarch
1 Tablespoon soy sauce
1/2 teaspoon ground ginger
1 cup plain yogurt

mint, chopped, for garnish

Noodles:
    1/4 cup vegetable oil
    2 onions, chopped
    1/2 cup water chestnuts, sliced
    1 Tablespoon soy sauce
    1 cup plain yogurt
    8 oz. egg noodles, cooked and drained

In a frying pan, sauté the onion and green pepper in the oil until golden, then add the pork cubes. Drain the pineapple, reserving the juice. Add the pineapple juice, brown sugar, vinegar, cornstarch, soy sauce, and ginger to the pork mixture; stir slowly until thickened. Add the pineapple chunks and yogurt and heat through.

Cook the noodles separately. Brown the onions in the oil, then stir in the water chestnuts, soy sauce, yogurt, and cooked noodles.

When ready to serve, put the sweet-and-sour pork over the noodles and garnish with fresh mint. Darjeeling tea makes a nice accompaniment.

PAKISTANI LEG OF LAMB

Pakistanis incorporate flavors from several cultures in their cuisine.
Serves 8.
6 to 8 lb. leg of lamb, with the bone left in

Marinade:
  1 cup soy sauce
  1/3 cup olive oil
  1 Tablespoon ground ginger
  3 cloves garlic, minced

Cucumber sauce:
  2 cucumbers, peeled, seeded, and chopped
  1 cup plain yogurt
  1 onion, chopped
  1 teaspoon ground cumin
  1/2 teaspoon salt
  1/4 teaspoon sugar
  1/4 teaspoon chili powder
  2 cups peas, cooked

Preheat oven to 325°. Marinate the lamb in a combination of the soy sauce, olive oil, ginger, and garlic for 1 whole day, covered and refrigerated; turn the meat occasionally. Roast until tender, about 2 to 2 1/2 hours.

Serve with *cucumber sauce*: combine the cucumbers, yogurt, onion, cumin, salt, sugar, chili powder, and peas.

Note: Greek leg of lamb (Arni Me Yaourti) is served with this spicy yogurt sauce: *1 cup plain yogurt*, 2 minced cloves garlic, 1 cup chopped tomatoes, 1/2 teaspoon flour, 1/2 teaspoon cinnamon, salt and pepper to taste.

PASTITSIO MAKARONIA ME FETA

Tempting layers of Greek macaroni with meat, zucchini, and feta cheese.
Serves 6 to 8.

1 lb. macaroni, cooked according to directions
1/4 cup oil
2 onions, chopped
2 cups zucchini, cut into 1/2-inch slices
2 lbs. round steak, ground
1 can (16 oz.) tomatoes, chopped
1 teaspoon salt
1/4 teaspoon pepper
1/4 teaspoon dried oregano
1/4 teaspoon ground cinnamon
1/4 teaspoon grated nutmeg
1 cup plain yogurt
2 egg yolks, lightly beaten

6 oz. feta cheese, crumbled, as a topping

Preheat oven to 375°. Cook the macaroni until tender but still firm; drain and set aside. Sauté the onion and zucchini in oil until wilted, then add the meat and cook until browned. Add the tomatoes, salt, pepper, oregano, cinnamon, and nutmeg; heat through.

Layer the macaroni and meat mixture alternately in a shallow 2-quart casserole dish. Stir the egg yolks into the yogurt, then pour over the top. Sprinkle feta cheese over the whole dish and bake until browned, about 45 minutes.

POLISH PIKE

The versatile horseradish sauce can be used as an accompaniment to many fish dishes.
Serves 6.

1 whole pike (about 3 lb.), cleaned
2 Tablespoons oil
2 onions, chopped
4 carrots, peeled and sliced
4 stalks of celery, sliced
2 Tablespoons fresh parsley, chopped
1 teaspoon salt
1/4 teaspoon pepper

Horseradish sauce:
  1 cup plain yogurt
  4 Tablespoons grated horseradish
  2 teaspoons lemon juice
  1/4 teaspoon sugar
  1/4 teaspoon salt

Preheat oven to 350°. Sauté the onions, carrots, and celery in the oil until the vegetables are tender. Add the parsley, salt, and pepper. In an ovenproof casserole, pour the sauce over the prepared fish and bake until tender, about 20 to 25 minutes.

When the fish is ready to serve, pass the *horseradish sauce*: mix together the yogurt, horseradish, lemon juice, sugar, and salt.

QUICHE LORRAINE FOR A CROWD

Makes 4 quiches, so invite all your friends for a French feast.
Serves 16 generously.

4 prepared 9-inch pie shells, partially baked and cooled
1 lb. bacon, cooked, drained, and crumbled
4 onions, chopped
1 1/2 lb. Swiss or Gruyere cheese, grated
1 cup Parmesan cheese, grated
1 dozen eggs, slightly beaten
1 cup plain yogurt
3 cups heavy cream
1 teaspoon salt
1/2 teaspoon pepper

grated nutmeg, as a topping

Preheat oven to 450°. Cook the bacon, then remove and crumble. Drain off most of the bacon fat, but cook the onions in about 2 Tablespoons of it until they become transparent. Sprinkle the cooked onions and bacon bits over the bottoms of the 4 prepared pie crusts, then add layers of the Swiss or Gruyere and Parmesan cheese. Combine the eggs, yogurt, cream, salt, and pepper; pour over the onion-bacon-cheese mixtures. Sprinkle the tops with nutmeg.

Bake the quiches for 15 minutes at 450°, then reduce the heat to 350° and bake for 10 to 15 minutes more, or until a knife inserted in the center comes out clean. Either serve right away or cool, cover with plastic wrap, refrigerate, and reheat later at 350° for 20 minutes.

RAVED-OVER ROMANIAN CHICKEN

"Pui Fragezit" is a smothered chicken.
Serves 4.

2 to 3 lb. chicken fryer
2 Tablespoons virgin olive oil
2 Tablespoons flour
1 cup plain yogurt
1 teaspoon salt
1/4 teaspoon pepper
2 Tablespoons fresh parsley, chopped
2 Tablespoons poppy seeds
1 cup water
2 teaspoons lemon juice

Accompaniments:
  lemon slices
  pasta

Brown the chicken pieces in the olive oil, then remove from the frying pan. Add the flour, then blend in the yogurt, salt, pepper, parsley, poppy seeds, and water. Bring to a boil.

Return the chicken and simmer covered, about 45 minutes, or until the chicken is tender. Stir in the lemon juice and serve over or beside pasta with thin slices of lemon.

ROGAN JAUSH

This famous Indian epicurean lamb dish translates to mean "Color-Passion Curry."
Serves 6.

2 lbs. boneless lamb, cubed
1/4 cup butter
2 onions, chopped
2 teaspoons ground coriander
1/2 teaspoon ground turmeric
1/2 teaspoon ground cumin
1/4 teaspoon ground ginger
1 cup plain yogurt
1 teaspoon salt
1/2 teaspoon paprika
2 tomatoes, peeled, seeded, and cut into chunks

Melt the butter in a frying pan and sauté the onions until golden brown. Add the lamb cubes, coriander, turmeric, cumin, and ginger; cook until the lamb is browned–about 15 minutes. Add the yogurt, salt, paprika, and tomato chunks. Cover and simmer gently until the lamb is tender, about 1 1/2 to 2 hours.

Serve hot with steamed rice.

SAUERBRATEN

Served with red cabbage and caraway noodles, this makes a perfect German meal.
Serves 6 to 8.

Marinade:
    1 cup cider vinegar
    1 cup dark red German wine
    2 onions, sliced
    1 carrot, peeled and sliced
    1 stalk celery, chopped
    2 Tablespoons granulated sugar
    1 Tablespoon salt
    4 whole cloves
    4 peppercorns

4 to 6 lb. beef pot roast, rump or chuck
2 Tablespoons vegetable oil
2 Tablespoons flour
1/2 cup ginger snaps, crushed
1 cup plain yogurt

fresh parsley, chopped, as a topping

In a large bowl combine the ingredients of the *marinade*: the vinegar, wine, onions, carrots, celery, sugar, salt, cloves, and peppercorns. Put the pot roast in the marinade, cover with a damp cloth, and refrigerate for 3 days, turning the meat occasionally.

When ready to cook the meat, dry with paper towels, put in a heavy kettle, and brown on all sides in the oil. Pour in the marinade and simmer, covered, for 2 1/2 to 3 hours, or until the meat is tender. Remove, sieve the vegetables, and skim off all the fat from the surface.

Make a paste from one-quarter of the drippings and 2 Tablespoons of flour; stir into the vegetable liquid, adding the gingersnaps and yogurt. Heat through but don't boil. Sprinkle with parsley. Spoon the gravy over the pot roast, then slice thin.

LE SOUFFLÉ SAUMON CANADIEN
(Canadian Salmon Soufflé)

A delightful salmon creation from our neighbor to the north.
Serves 4.

4 Tablespoons butter or margarine
4 Tablespoons flour
1/4 teaspoon salt
dash of pepper
dash of grated nutmeg
1 cup plain yogurt
4 eggs, separated
1 lb. salmon, fresh or canned

Garnishes:
   parsley, chopped
   hard-cooked eggs, sliced

Preheat oven to 350°. Make a white sauce by melting the butter in a small saucepan, stirring in the flour, salt, pepper, and nutmeg, and then slowly adding the yogurt; heat until thickened. Blend in the egg yolks and salmon.

In a separate bowl, beat the egg whites until stiff, then fold into the salmon mixture. Carefully pour into a buttered 1-quart soufflé dish and bake until puffed, about 45 minutes. Serve immediately, garnishing with parsley and hard-cooked eggs.

STUFFED CZECH CHOPS

"Veprove s Krenem"–Czechoslovakian porkchops with horseradish.
Serves 6.

1/4 cup butter or margarine
1 onion, chopped
1/2 cup mushrooms, sliced
2 slices dark bread, crumbled (I prefer pumpernickel)
1 egg, slightly beaten
1/2 teaspoon salt
1/4 teaspoon pepper
1/4 teaspoon Hungarian paprika
6 double pork chops, split for pockets
1/2 cup chicken stock
1/4 cup cognac
1 cup plain yogurt
1/4 cup prepared horseradish

paprika, for garnish

In a frying pan, lightly cook the onion in the butter, then add the mushrooms and breadcrumbs; toss until brown. Beat the egg slightly, then add the salt, pepper, and paprika. Combine the 2 mixtures and stuff into the pork chops, fastening with toothpicks where necessary.

In the same frying pan, adding more butter if needed, brown the pork chops on both sides. Stir in the chicken stock and simmer until the chops are cooked through, about 45 minutes.

Pour on the cognac, ignite, and remove the pork chops to a serving platter. Stir in the yogurt and horseradish, heat through, and use as a sauce. Garnish lightly with paprika. Serve with caraway noodles.

SWISS SPAGHETTI-VEAL CASSEROLE

"Emince de Veau avec Yaourt" is perfect served with pasta.
Serves 4 to 6.

4 Tablespoons butter or margarine, melted
2 lbs. veal scallops, thinly sliced
4 Tablespoons flour
1 cup beef stock
1/2 cup dry white wine
1 cup Gruyere cheese, grated
1 cup plain yogurt
1/2 teaspoon salt
dash of pepper

1 lb. spaghetti noodles, cooked

Quickly brown the veal slices in the melted butter, then remove from the frying pan. Stir in the flour, beef stock, and wine; heat until thickened. Add half the grated cheese. Stir in the yogurt, salt, pepper, and veal slices.

Serve over cooked pasta, with the remaining cheese sprinkled on top.

THAI CHICKEN CURRY

This dish is a favorite of our Siamese cats!
Serves 6 to 8.

1/4 cup peanut oil
2 onions, chopped
1 clove garlic, minced
1/4 cup flour
1 Tablespoon curry powder
1 teaspoon salt
1 teaspoon sugar
1/4 teaspoon ground ginger
dash of cayenne pepper
2 cups chicken broth
4 cups chicken, cooked and cubed
1 cup plain yogurt
1 teaspoon lemon juice

rice, cooked

Accompaniments: chutney, pineapple chunks, coconut, chopped egg, bacon bits, pickles, or raisins.

In a frying pan, sauté the onions and garlic in the peanut oil until slightly browned. Add the flour, curry, salt, sugar, ginger, and cayenne pepper; heat until thickened. Stir in the chicken broth and cook for 5 to 10 minutes.

Add the chicken cubes, yogurt, and lemon juice. Heat until warmed through, then serve over rice with these *accompaniments*: fruit chutney, pineapple chunks, coconut, chopped egg, bacon bits, pickles, raisins, etc.

WEST AFRICAN BEEF-LIVER KABOBS

Hot and spicy "Kyinkyinga" are traditionally served with rice.
Serves 4 to 6.

1 lb. beef chuck, tip or roast
1 lb. beef liver

Marinade:
  1 cup plain yogurt
  4 Tablespoons lemon juice
  2 Tablespoons oil
  1 cup onion, chopped
  1 clove garlic, minced
  1/4 teaspoon ground ginger
  1/8 teaspoon cayenne pepper

rice, boiled or steamed, as an accompaniment

Cut the beef chuck and the beef liver into 1-inch cubes. Cover with the *marinade*: mix together the yogurt, lemon juice, oil, onion, garlic, ginger, and cayenne pepper; refrigerate for several hours, turning occasionally.

Thread the beef cubes onto kabob skewers and broil until browned, about 10 minutes. Serve with rice and pass the extra marinade sauce.

YAM-HAM PIE

Residents of the Dominican Republic call this creation "Pastel de Mapueyes."
Serves 6.

4 yams, peeled and thinly sliced
3 cups cooked ham, cubed
2 Tablespoons flour
2 Tablespoons brown sugar
1/2 teaspoon dry mustard
1 cup plain yogurt
12 green olives, halved
2 Tablespoons butter or margarine

Preheat oven to 400°. Prepare the yams, then arrange them in a 9-inch round buttered pie pan. Top with the ham cubes.

Combine the flour, brown sugar, dry mustard, and yogurt; pour over the yam-ham mixture. Cover with the olive halves, dot with butter, and bake until heated through, about 45 minutes.

YANKEE RED FLANNEL HASH

Yankee ingenuity has always found a hearty use for leftovers.
Serves 6 to 8.

leftover cooked corned beef (about 4 cups)
leftover potatoes from "New England boiled dinner" (4 to 6)
leftover cooked carrots (2 cups)
leftover cooked cabbage (about half a head)
leftover cooked onions (4 to 6)
leftover cooked turnips (4 to 6)
leftover cooked beets (2 cups) or 1 lb. canned beets, drained
1 cup plain yogurt
1 Tablespoon Worcestershire sauce
1/4 cup catsup
1 teaspoon salt
1/4 teaspoon pepper
dash of hot pepper sauce
1/4 cup vegetable oil

Using either a food processor or meat grinder, cut up the corned beef, potatoes, carrots, cabbage, onions, turnips, and beets and put into a large bowl. In a separate bowl, mix together the yogurt, Worcestershire sauce, catsup, salt, pepper, and hot pepper sauce; combine the two mixtures.

Heat the oil in a large frying pan and fry hash until crusty on both sides. Or, if you want a less crispy hash, heat some or all of it in a 350° oven for 30 to 40 minutes.

Leftovers can be kept refrigerated at least a week. It can be served under poached eggs for breakfast or as a whole meal in itself. This recipe is a good first "people food" for baby.

Note: This recipe won a New England cookbook contest!

YUGOSLAVIAN CHICKEN AND ASPARAGUS

Called "Pilece Meso sa Sparglama," this is sensational when asparagus is in season.
Serves 4 to 6.

2 lbs. fresh asparagus, trimmed (or 2 10-oz. packages, cooked)
1/4 cup olive oil
4 cups chicken, cooked and cubed
1 teaspoon salt
1/4 teaspoon pepper
1/4 teaspoon dried thyme
dash of grated nutmeg
1 cup plain yogurt
2 eggs, beaten

Preheat oven to 375°. Cook the asparagus until just tender (microwaving makes it easy!); drain, then arrange in the bottom of a lightly greased, oblong casserole. Briefly heat the chicken in the oil, then add the salt, pepper, thyme, and nutmeg. Pour over the asparagus.

Blend together the yogurt and eggs, then pour over the chicken-asparagus mixture. Bake for about 20 minutes, or until the top is golden brown.

Sweets For All

All love is sweet,
Given or returned. Common as light is love.

–Shelley's *Prometheus Unbound*

What greater way to show your love than through creating and serving deliriously delectable desserts?

Food and drink for the Olympic gods, *Amazing Ambrosia* is a Greek version of fruit topped with an orange-flavored liqueur yogurt sauce, coconut or crushed macaroons, and cherries.

Keep your cookie jar stocked with *Arabian Sesame Snacks*, a nutritious after-school snack for the kids. *Carob Cookies* are a chocolate substitute that makes 5 to 6 dozen delights. We love *Gingerbread Men*, and once used them as the main decor for an "Old-Fashioned Wassail" party we had at Christmastime.

Armenian Lemon Bars (Madzoonov Gargantag) are especially nice following a fish dish, and *Brazilian Nutty Nutmeg Nuggets* taste terrific with ice cream. "Tourta Me Yaourti Kai Stafides," *Spicy Greek Squares*, are sweet yogurt-raisin treats. And *Dutch Dessert Delight* is sure to delight the birthday girl or boy!

Puddings take on a whole new taste with yogurt, such as *Barbados Banana Pudding*. An elegant egg custard, *Bavarian Apricot Cream* can be made for many different friends. Jewish families serve *Israeli Sweet Noodle Kugel* during the holiday season, while *Norwegian Rhubarb Pudding* (Rabarbragrot) signals springtime in Scandinavia. *Spanish St. Joseph's Day* is celebrated by a festival in Valencia, Spain on March 19th where plenty of custard is enjoyed. *Singapore Tapioca* is an option with Oriental fare, and *Royal Raspberry Pudding* befits the King and Queen anytime.

Pies also lend themselves nicely to yogurt fillings. Good examples are *Australian Ribbon Pie*, a lovely two-layered lemon-coconut pie, and *Frozen Finnish Fruit Pie*, which can be prepared ahead of time. *Peach Pie with Gingersnap Crust* is a winner from the West Indies. "Calabaza," *Trinidadian Pumpkin Pie*, is one of our favorites, and *Southern Pecan Pie* is worth breaking a diet for.

Belgian Berries, in any combination, are enhanced when topped with a sauce of yogurt, honey, and cinnamon. Another fantastic fruit concoction is *Brandied Polynesian Pineapple*, which adds luscious liqueurs to your next luau. *Grapes of No Wrath* take only seconds to prepare, but are simply scrumptious.

Cheesecake is everyone's favorite; be sure to try *Cuban Cheesecake*, made with crushed pineapple, or *Italian Apple Cheesecake*, (Torta di Mele).

On November 5th be sure to make up a batch of *Guy Fawkes Fingers*, named for the conspirator who failed to blow up England's Houses of Parliament in 1605. The Americans commemorate their clipper ships with *Heritage Hermits*.

Greek Apple Crisp can be served hot or cold. An elegant and unusual treat from Turkey is *Turkish Strawberry-Pistachio Dessert*, a yogurt base topped with fruit, nuts, and whipped cream.

The *Jamaican Gingerbread House* makes a nice housewarming gift; shape the cooked batter into a house, frost, and decorate. *Yogurt Popsicles* are another fun recipe to have on hand; directions for Frozen Fruit popsicles and Nutter-Butter Pops are included, but you are encouraged to experiment with other combinations. Or, you might want to revel with rum–try the *Dorado Beach Daiquiri Ice Cream*.

And when you're really ready for a binge, you can still do it in a healthy manner, with a *Banana Split* covered with a yogurt-brown sugar mixture, then topped with a number of extra goodies.

SWEETS FOR ALL

AMAZING AMBROSIA

Greek fruit ambrosia was food and drink for the Olympic gods.
Serves 6 to 8.

4 peaches, peeled and sliced
2 bananas, peeled and sliced
1 apple, peeled, cored, and diced
1 can (8 oz.) crushed pineapple, drained
1 cup grapes, halved
1 cup blueberries or sliced strawberries
2 grapefruits, sectioned
2 oranges, sectioned
1/2 cup figs or dates, chopped
2 pears, unpeeled and diced

Sauce:
   1/4 cup orange-flavored liqueur
   1 cup plain yogurt
   2 Tablespoons honey

2 cups coconut or 1/2 lb. macaroons, crushed, for topping
cherries, for decoration

Prepare and combine the peaches, bananas, apple, pineapple, grapes, blueberries or strawberries, grapefruit, oranges, figs or dates, and pears in an attractive serving bowl.

Stir the orange-flavored liqueur and honey into the yogurt and pour over the fruit. Sprinkle with coconut or crushed macaroons and decorate each portion with a cherry.

ARABIAN SESAME SNACKS

Ali Baba will yell "Open Sesame" to your cookie jar!
Makes about 4 dozen cookies.

1/2 cup butter or margarine
1/2 cup shortening
1 cup sugar
1 egg, slightly beaten
1 cup plain yogurt
1 teaspoon vanilla extract
1/4 teaspoon grated lemon rind
3 cups flour
1 teaspoon baking soda
1 teaspoon salt
sesame seeds, for topping

Preheat oven to 375°. Cream together the butter, shortening, and sugar. Add the egg and beat until the batter is thick. In a separate bowl, stir the vanilla and lemon rind into the yogurt. Sift the flour with the baking soda and salt. Add the flour mixture and the yogurt alternately to the butter-sugar mixture, beginning and ending with the flour. Refrigerate for several hours, until firm.

Drop teaspoonfuls of the batter onto a greased cookie sheet. Flatten and top with sesame seeds. Bake until lightly browned, about 10 to 12 minutes.

ARMENIAN LEMON BARS

"Madzoonov Gargantag" are delightful following a fish dish.
Makes 9 to 12 squares.

1/2 cup butter or margarine
1 1/2 cups sugar
1 cup plain yogurt
3 eggs, beaten
1 Tablespoon lemon juice
1 teaspoon lemon rind, grated
2 1/2 cups flour
1/2 teaspoon baking soda

confectioner's sugar, for topping

Preheat oven to 350°. Cream together the butter and sugar in a large bowl; add the yogurt and mix well. Beat in the eggs, lemon juice, and lemon rind. Sift the flour with the baking soda, then fold into the lemon mixture.

Pour the batter into a greased 9-inch square baking pan and bake until the bars test done, about 35 to 40 minutes. When cool, sprinkle with confectioner's sugar and cut into bite-sized bars.

AUSTRALIAN RIBBON PIE

This is a lovely, two-layered lemon-coconut pie.
Serves 6 to 8.

1 prepared 9-inch pie shell, baked
1/3 cup butter or margarine
1 cup granulated sugar
3 eggs, beaten
juice of 2 lemons
grated rind of 2 lemons, divided
2 Tablespoons granulated sugar
1 Tablespoon cornstarch
1 cup plain yogurt
1/4 teaspoon salt
1 teaspoon vanilla extract
1/4 cup flaked coconut

In a small saucepan, melt the butter in the top of a double boiler and cream in the sugar. Stir in the beaten eggs, lemon juice, and half of the lemon rind. Cook over boiling water, stirring constantly, until thickened; pour into the baked pie shell and let cool.

Combine the cornstarch, sugar, yogurt, salt, and vanilla. Cook and stir until thickened; add the coconut. Pour over the lemon mixture in the pie shell and garnish with the rest of the lemon rind. Refrigerate until well chilled. When ready to serve, cut into wedges.

BANANA SPLIT

A healthy version of the ultimate splurge.
Serves 1.

1 banana, peeled and split lengthwise
1 Tablespoon brown sugar
1 cup plain yogurt

Toppings:
   granola
   nuts
   seeds
   crushed toffee crunch
   chocolate bits or sprinkles
   candies

Place the split banana on a serving dish. Blend together the brown sugar and yogurt and pour over the banana.

Top with any of the above suggestions, or make your own combinations. Enjoy!

Note: This dessert is a nice variation for birthday parties, regardless of age.

BARBADOS BANANA PUDDING

West Indians call bananas the "fruit of the wise men."
Serves 6 to 8.

4 bananas, peeled and cut into chunks
juice of 2 limes
2 eggs, slightly beaten
6 slices of leftover bread, finely crumbled
1 cup plain yogurt
1/2 cup granulated sugar
1/4 cup butter, melted
1/2 teaspoon vanilla
1/4 teaspoon ground cinnamon

Preheat oven to 375°. Mash the bananas and mix with the lime juice. Beat in the eggs, then add the breadcrumbs, yogurt, sugar, melted butter, vanilla, and cinnamon, mixing well.

Pour the mixture into an 8-inch square baking dish and bake for about 1 hour. Remove the pudding and let it cool a bit before serving.

You can decorate the top with extra banana slices.

BAVARIAN APRICOT CREAM

This elegant apricot egg custard can be made with many different fruits.
Serves 6.

1 cup dried apricots
water
1 envelope unflavored gelatin
1/4 cup fruit juice (any flavor)
1/4 cup honey
2 eggs, separated
1 teaspoon lemon juice
1 teaspoon lemon rind, grated
1 cup plain yogurt
1 Tablespoon apricot brandy (optional)

whipped cream, for topping

Soak the dried apricots in enough water to cover until softened. Dissolve the unflavored gelatin in the fruit juice; set aside. Drain the apricots, reserving 1 cup of the juice, and chop them into small pieces. Put the chopped apricots into a saucepan with the honey, egg yolks, lemon juice, lemon rind, and the 1 cup of reserved apricot water. Bring to a boil, then remove from the heat and add the dissolved gelatin. Chill the apricot mixture until it becomes the consistency of egg whites.

Beat the egg whites until stiff; fold into the apricot mixture. Fold in the yogurt and apricot brandy, pour into a 1-quart mold, and chill until firm, about 3 to 4 hours. Serve with whipped cream, if desired.

BELGIAN BERRIES

*Belgium has miles of hothouses in which various berries are culti-
vated.*
Serves 4.

1 cup strawberries
1 cup raspberries
1 cup blueberries
1 cup lingonberries
1 cup blackberries
1 cup elderberries
1 cup gooseberries

Sauce:
  1 cup plain yogurt
  2 Tablespoons honey
  1/2 teaspoon ground cinnamon

Either combine the berries in a bowl or put them in individual
dishes and let people choose between them.

For the *sauce*: mix together the yogurt, honey, and cinnamon and
serve on top of the berries.

BRANDIED POLYNESIAN PINEAPPLE

Add luscious liqueurs to your next luau.
Makes about 1 cup.

1 fresh pineapple, peeled, cored, and cut into slices or chunks

Sauce:
> *1 cup plain yogurt*
> 2 Tablespoons fruit-flavored brandy
> 1 Tablespoon lime juice

fresh mint, chopped
coconut

Blend together the yogurt, brandy, and lime juice. Pour over the pineapple slices or chunks and arrange in a serving dish.

Decorate with mint and/or coconut.

BRAZILIAN NUTTY NUTMEG NUGGETS

We enjoyed these Brazil nut squares while visiting the Amazon River recently.
Serves 9.

2 cups flour
1 1/2 cups brown sugar
1/2 cup shortening
1 egg, slightly beaten
1 teaspoon grated nutmeg
1 cup plain yogurt
1 teaspoon baking soda
1 cup Brazil nuts, chopped

Preheat oven to 350°. Blend the flour, brown sugar, and shortening with a pastry blender until it forms a coarse, crumby mixture. Put half the crumbs into a greased 9-inch square pan. To the remaining crumbs add the egg and nutmeg. Dissolve the baking soda in the yogurt, fold into the remaining crumb mixture, and pour into the pan.

Sprinkle the mixture with the chopped Brazil nuts. Bake until the nuggets test done, about 35 to 40 minutes. Cut into squares and serve with ice cream.

CAROB COOKIES

Known as "St. John's bread," carob is a healthy, low fat chocolate substitute.
Makes 5 to 6 dozen.

2 eggs, slightly beaten
1/2 cup safflower oil
1 cup plain yogurt
2 cups sugar
1 teaspoon vanilla extract
2 cups flour
1/2 cup wheat germ
1 1/2 cups rolled oats
1/2 cup carob powder
1/2 cup walnuts, chopped

Preheat oven to 375°. In a bowl, beat the eggs into the oil, yogurt, sugar, and vanilla. Sift the flour with the wheat germ, oats, and carob and combine with the yogurt-sugar mixture. Stir in the chopped walnuts.

Drop the carob batter by rounded teaspoonfuls onto greased cookie sheets and bake until browned, about 12 to 15 minutes.

CUBAN CHEESECAKE

Pineapple, frequently included in Cuban cuisine, really makes this cheesecake.
Serves 6 to 8.

Crust:
    1 cup graham crackers, crushed
    2 Tablespoons butter or margarine, melted
    1/2 cup pecans, chopped

8 oz. can crushed pineapple
1 envelope unflavored gelatin
1/4 cup reserved pineapple juice
8 oz. cream cheese, softened
1 cup plain yogurt
1/4 cup sugar
1/2 teaspoon vanilla extract

Prepare the crust: combine the crushed graham cracker crumbs, melted butter, and pecans. Press into an 8-inch pie pan; set aside.

For the topping, drain the pineapple, reserving 1/4 cup of the liquid. Sprinkle the gelatin over the reserved juice and let it soften. When the gelatin has dissolved, put it in a blender with the cream cheese, yogurt, sugar, and vanilla; blend thoroughly. Stir in the pineapple. Pour into the graham cracker crust and chill until set, several hours.

DORADO BEACH DAIQUIRI ICE CREAM

A daiquiri delight from the hotel where we honeymooned in Puerto Rico.
Serves 4.

Fruit:
  1 cup bananas, mashed
  2 cups peaches, chopped

1/2 cup light corn syrup
4 Tablespoons sugar
1/4 teaspoon lemon or lime rind, grated
1 Tablespoon lemon or lime juice
2 cups plain yogurt
2 Tablespoons Puerto Rican rum

Combine the ingredients in this order in an electric blender: fruit, corn syrup, sugar, grated lemon or lime rind, lemon or lime juice, and yogurt. Blend on medium speed until well mixed. Pour into a 9 × 5 inch loaf pan, cover with aluminum foil, and freeze until firm, about 4 hours.

Put the mixture back in the blender with the rum and blend 1 more minute at medium speed. Return to the freezer until frozen.

DUTCH DESSERT DELIGHT

"Verjaardag Koek" means birthday cake.
Serves 9-12.

1 1/4 cups sugar
3/4 cup butter or margarine
4 eggs
3 oz. semi-sweet chocolate, melted and cooled
rind of 1 lemon, grated
2 1/2 cups flour
1 teaspoon baking soda
1/4 teaspoon salt
1 cup plain yogurt

Frosting:
   1 cup butter or margarine
   1 cup confectioner's sugar
   1/4 cup Dutch cocoa

Decoration:
   1 1/2 cups whipped cream
   2 Tablespoons sugar
   1 teaspoon vanilla (or Dutch liqueur)
   chocolate curls

Preheat oven to 350°. Cream the sugar and butter until light and fluffy. Add the eggs one at a time; blend. Add the chocolate and the lemon rind; blend.

Sift the flour with the baking soda and salt. Add the dry ingredients and the yogurt alternately to the sugar-butter mixture. Pour into 2 9-inch round pans and bake for about 30 minutes.

When cooled, cover with the chocolate *frosting:* cream the butter with the confectioner's sugar and Dutch cocoa. Decorate with the whipped cream (made with whipping cream, sugar, and vanilla or liqueur). Place chocolate curls around the top.

FROZEN FINNISH FRUIT PIE

Finland, we found, abounds with fabulous fresh fruit–your choice.
Serves 6 to 8.

Crust:
    4 Tablespoons butter or margarine, melted
    6 oz. package chocolate bits
    2 cups crisp rice cereal

2 Tablespoons sugar
4 Tablespoons of your favorite fruit preserves
4 cups plain yogurt

fruit or chocolate sprinkles, for decoration

    For the base, melt the butter and chocolate bits together; stir in the cereal. Mix, then use to line a 9-inch pie pan.
    Stir the sugar into the fruit preserves and combine with the yogurt. Pour into the prepared pie crust and freeze until firm, several hours.
    When ready to serve, top with fresh fruit or chocolate sprinkles.

GINGERBREAD MEN

These cookies were our main decor for an "Old-Fashioned Wassail" party.

Makes about 4 dozen, depending on the size of your cookie cutter.

1 cup butter or margarine
1 cup molasses
1 cup honey
1 cup plain yogurt
6 cups flour
4 teaspoons double-acting baking powder
2 teaspoons ground ginger
1 teaspoon baking soda
1 teaspoon salt
1 teaspoon grated lemon rind

Decoration:
   raisins
   cinnamon drops
   "bullets"
   decorative frosting

Preheat oven to 350°. Melt the butter in a saucepan, then mix with the molasses and honey. Stir in the yogurt. In a separate bowl, sift the flour with the baking powder, ginger, baking soda, and salt. Blend the two mixtures together, then beat in the grated lemon rind. Chill for several hours, until the dough is firm.

Roll the dough out on a floured surface to about 1/4 inch. Cut with a cookie cutter shaped like a gingerbread man. Put on ungreased cookie sheets and bake until lightly puffed and browned, about 8 to 10 minutes. Cool completely, then decorate with raisins for the eyes, cinnamon drops for the mouth, and "bullets" for buttons. Use decorative frosting also, if you prefer.

GRAPES OF NO WRATH

My apologies to Steinbeck, but these are so simple and refreshingly good.
Serves 4.

1 lb. seedless grapes
1 cup plain yogurt

Topping:
    1/2 cup brown sugar
    ground cinnamon

Mix together the grapes and the yogurt. Put in individual serving dishes.
    Top with the brown sugar and sprinkle with cinnamon.

GREEK APPLE CRISP

Greek desserts are justly famous for their flavors.
Serves 6 to 8.

6 to 8 baking apples, pared and sliced
1 cup brown sugar
1 cup flour
1/4 cup butter
1/4 teaspoon salt
1/4 teaspoon lemon rind, grated

Sauce:
   1 cup plain yogurt
   1/4 cup sugar
   2 teaspoons ground cinnamon
   pinch of salt

Preheat oven to 350°. Grease a 9-inch square baking dish and line with apples until it is full. Mix together the brown sugar, flour, butter, salt, and lemon rind with a pastry blender. Spread over the apples, then bake until the top is browned, about 40 to 45 minutes.

Serve hot or cold with a sauce made by mixing together the yogurt, sugar, cinnamon, and salt.

Note: This is a Fall favorite at our house.

GUY FAWKES FINGERS

*Served in England on Bonfire Night, November 5th, as Fawkes is
burned in effigy.*
Makes 9 to 12 squares.

1/2 cup molasses
1/3 cup butter or margarine
1/2 cup brown sugar
1/2 teaspoon baking soda
1 egg, slightly beaten
2 cups flour, sifted
2 teaspoons ground ginger
1 teaspoon double-acting baking powder
1/2 teaspoon salt
1 cup plain yogurt
grated rind of 1 lemon

Icing:
    2 cups confectioner's sugar
    2 to 4 Tablespoons lemon juice
    1/2 teaspoon vanilla extract

Preheat oven to 325°. Heat the molasses and butter together in a
saucepan until melted. Add the brown sugar and baking soda, then
let cool. Beat in the egg. Resift the flour with the ginger, baking
powder, and salt in a large bowl. Make a well in the center of the
flour mixture and pour in the molasses mixture and yogurt, mixing
the batter until smooth. Add the lemon rind, beat well, and pour into
a buttered and floured 9-inch cake pan.

Bake for about 15 to 20 minutes, or until the cake springs back
when touched. Remove and let cool for at least 10 minutes before
covering with the *icing*: mix together the confectioner's sugar and
lemon juice to a thin consistency, then add the vanilla extract. When
cool, cut into finger shapes to symbolize the conspirator who failed
to blow up the Houses of Parliament in 1605.

HERITAGE HERMITS

These traditional American treasures arrived with the clipper ships.
Makes about 36 per baking sheet.

2 eggs, slightly beaten
1 cup shortening
2 cups brown sugar
3 cups white flour
3 cups whole wheat flour
2 teaspoons ground cinnamon
2 teaspoons ground cloves
2 teaspoons baking soda
1 teaspoon grated nutmeg
1 teaspoon salt
1 cup plain yogurt
1 cup molasses
1 cup raisins
1/2 cup walnuts, chopped (optional)

Preheat oven to 350°. To the slightly beaten eggs add the shortening and brown sugar together in a bowl; blend well. Sift the white and wheat flour with the cinnamon, cloves, baking soda, nutmeg, and salt. Add the dry ingredients and the yogurt and molasses alternately to the shortening-sugar mixture. Stir in the raisins and walnuts.

Spread the batter onto 2 large, greased cookie sheets. Bake for about 20 to 25 minutes, or until browned, cutting into squares when done.

ISRAELI SWEET NOODLE KUGEL

Many Jewish families serve variations of this on their holidays.
Serves 4 to 6.

8 oz. package egg noodles, medium-wide
2 eggs, separated
1 cup uncreamed cottage or pot cheese
1 cup plain yogurt
1/2 cup sugar
1 teaspoon ground cinnamon
1/4 teaspoon grated nutmeg
dash of salt
1/2 cup raisins (optional)
1/4 cup graham cracker crumbs

Preheat oven to 350°. Cook the noodles according to the package directions; drain. Beat the egg whites until stiff, then set aside. Add the egg yolks to the noodles, then blend in the cottage or pot cheese, egg whites, yogurt, sugar, cinnamon, nutmeg, and salt. Fold in the raisins.

Transfer the mixture to a buttered 1 1/2-quart casserole dish, top with graham cracker crumbs, and bake until brown and crusty on the top, about 45 minutes.

ITALIAN APPLE CHEESECAKE

Called "Torta di Mele," this cheesecake is a special fruit treat.
Serves 8.

1 8-inch prepared pie shell
l egg yolk, lightly beaten
2 Tablespoons Italian sherry
2 apples, cored, peeled, and pared
2 eggs
1/2 cup granulated sugar
2 Tablespoons flour
1/4 teaspoon salt
2 teaspoons lemon peel, grated
1 cup plain yogurt
1 cup ricotta cheese
1 Tablespoon candied orange peel, chopped
1/4 cup raisins, golden seedless

Preheat oven to 350°. Slightly beat the egg yolk, combine with the Italian sherry, and brush on the prepared pie crust. Cut the apples into wedges and arrange, slightly overlapping, on top of the pastry.

Beat the 2 eggs and sugar at high speed in an electric mixer; reduce the speed and add the flour, salt, lemon peel, yogurt, ricotta, orange peel, and raisins. Pour this mixture over the apples and bake for 60 to 75 minutes, or until golden. Let stand for 10 minutes before serving, or cool completely.

JAMAICAN GINGERBREAD HOUSE

This ginger treat makes a welcome housewarming gift.
Serves 6 to 8.

1/2 cup butter or margarine
1 cup molasses
1 cup plain yogurt
2 1/3 cups flour
1 teaspoon Jamaican ground ginger
1 teaspoon ground cinnamon
3/4 teaspoon baking soda
1/4 teaspoon salt
1/4 teaspoon ground cloves

Preheat oven to 350°. Melt the butter in a saucepan, then add the molasses and bring to a boil. Cool slightly, then add the yogurt. Stir the flour, ginger, cinnamon, baking soda, salt, and cloves into the mixture.

Pour the batter into a greased and floured 9-inch square cake pan, or preferably one already shaped like a house. Bake until the cake tests done, about 40 minutes.

Cut away the top of the cake to make a distinctive roof and chimney and proceed to decorate with colored frostings and candies.

NORWEGIAN RHUBARB PUDDING

Refreshing "Rabarbragrot" signals springtime in Scandinavia.
Serves 4.

1 1/2 lbs. rhubarb, trimmed and cut into 1/2-inch slices (4 cups)
1 1/2 cups water
3/4 cup sugar
3 Tablespoons cornstarch
1/4 cup cold water
1 teaspoon vanilla extract

Sauce:
1 cup plain yogurt
2 Tablespoons sugar
1/2 teaspoon vanilla extract
1/2 teaspoon ground cinnamon
1/4 teaspoon ground nutmeg

Prepare the rhubarb, then combine with the water and sugar in a saucepan. Cook until soft, about 10 minutes. Blend the cornstarch with the cold water to make a smooth liquid; stir slowly into the rhubarb mixture. Cook over low heat, stirring constantly, until thickened and clear. Add the vanilla, cover, and chill.

When ready to serve, add dollops of the *sauce*: blend together the yogurt, sugar, vanilla, cinnamon, and nutmeg.

PEACH PIE WITH GINGERSNAP CRUST

This pretty pie is a winner from the West Indies.
Serves 6.

Gingersnap piecrust:
    1 cup gingersnaps, crumbled (about 16 to 20)
    2 Tablespoons butter or margarine, melted

Topping:
    1 can (8 1/2 oz.) sliced peaches, drained and chopped
    2 Tablespoons honey or sugar
    1 teaspoon lemon juice
    1 envelope unflavored gelatin
    1 cup plain yogurt
    2 egg whites, beaten until stiff

Preheat oven to 350°. Prepare the *pie crust:* blend together the crumbled gingersnaps and melted butter, then press into an 8-inch pie pan. Bake for 8 to 10 minutes, or until browned; cool.

For the topping, drain the peaches, keeping out 6 slices for decoration. Chop and beat the rest of the peaches with the honey or sugar and lemon juice. Sprinkle the unflavored gelatin in the yogurt; stir until dissolved, then add to the peach mixture. Fold in the stiffly beaten egg whites. Turn the chiffon into the prepared gingersnap piecrust, garnish with the extra peach slices, and refrigerate until firm, about 3 to 4 hours.

ROYAL RASPBERRY PUDDING

This proud pudding is fit for Buckingham Palace!
Serves 6.

Sauce:
    1 10-oz. package frozen raspberries, thawed
    1/2 cup currant jelly
    1 Tablespoon water
    1 teaspoon cornstarch

8 buttered slices of bread, crusts removed
2 Tablespoons raisins, golden seedless
2 Tablespoons currants
2 Tablespoons glacéed mixed fruits
4 eggs
1/4 cup granulated sugar
1 teaspoon grated lemon peel
1/2 teaspoon vanilla
dash of salt
dash of nutmeg
1 cup plain yogurt
1 cup milk

Preheat oven to 350°. In a saucepan, heat together the thawed raspberries and currant jelly; bring to a boil, then add the water and cornstarch and cook, stirring constantly, until thickened. Strain and refrigerate.

Arrange 4 of the buttered bread slices in the bottom of a lightly buttered casserole dish. Sprinkle with a combination of half of the raisins, currants, and glacéed fruits; repeat this process. Now beat the eggs with the sugar, lemon peel, vanilla, salt, and nutmeg. Stir in the yogurt and milk, then pour over the raisin mixture. Let the mixture stand for 30 minutes, then bake for 1 hour, covered for the first half and then uncovered until golden.

Serve at once, topping with the raspberry sauce.

SINGAPORE TAPIOCA

Tapioca is a delightful dessert after an Oriental meal.
Serves 4 to 6.

4 apples, cored, peeled, and sliced (Spencers are spectacular)
2 Tablespoons water
1 cup tapioca
4 cups milk
1/2 cup sugar
rind of 1 lemon, grated
4 eggs, separated
1 can (8 oz.) crushed pineapple, drained
1 cup plain yogurt
1 teaspoon grated nutmeg
1 teaspoon ground cinnamon

whipped cream, for topping

Preheat oven to 375°. Put the apples and water in a 9-inch square pan and bake for about 1/2 hour. When softened, put through a sieve and set aside.

Place the tapioca, milk, sugar, and lemon rind in the top of a double boiler and cook over medium heat until tender, about 20 minutes. Add the apple pulp, egg yolks, pineapple, yogurt, and spices; mix well. Beat the egg whites until stiff, then fold into the tapioca. Pour the mixture into a buttered 2-quart casserole and bake at 325° for 30 to 35 minutes. Serve hot with sweetened, whipped cream.

SOUTHERN PECAN PIE

A custard-like version of the classic diet-breaker.
Serves 6.

2 eggs, slightly beaten
1 cup plain yogurt
1 cup brown sugar
1 teaspoon flour
1/2 teaspoon lemon rind, grated
1/4 teaspoon ground cinnamon
1/4 teaspoon ground cloves
1 cup pecans, chopped

9-inch pie crust, unbaked

whipped cream, for topping

Preheat oven to 450°. Beat the eggs, then add the yogurt, brown sugar, and grated lemon rind. Mix together the flour, cinnamon, and cloves. Add 2 Tablespoons of the yogurt mixture to the flour, then stir the flour and spices into the yogurt.

Sprinkle the pecans over the pie shell, then pour the batter over them.

Put the pie in the 450° oven, but turn the heat down right away to 325°. Bake until firm, about 40 minutes. Serve hot or cold, with or without whipped cream.

SPANISH ST. JOSEPH'S DAY CUSTARD
(Crema de San Jose)

St. Joseph's Day is a festival celebrated in Valencia, Spain on March 19th.
Serves 6.

4 egg yolks
1/2 cup granulated sugar
1 cup plain yogurt
3 Tablespoons cornstarch
3 cups milk
1 cinnamon stick
grated rind of 1 lemon
4 Tablespoons granulated sugar
1 teaspoon ground cinnamon

ladyfingers
whipped cream, for topping

Beat the egg yolks with the 1/2 cup granulated sugar and set aside. Dissolve the cornstarch in the yogurt. Heat the milk to the simmering point and add the cinnamon stick and grated lemon rind, then strain the hot milk into the bowl with the egg yolks and sugar. Add the yogurt and stir to combine. Return the whole mixture to the stove and cook, over low heat, until it is at the scalding point–but don't let it boil.

Remove the mixture from the heat and pour into a 1-quart casserole dish. Chill for about 4 hours, or until firm. Sprinkle the remaining sugar and cinnamon over the top of the custard to form a thin layer. Place under a preheated broiler for 5 minutes to carmelize the cinnamon-sugar and form a crust. Serve as is, or with ladyfingers and whipped cream.

SPICY GREEK SQUARES

"Tourta Me Yaourti Kai Stafides" are sweet yogurt-raisin treats.
Makes 9 to 12 squares.

1/2 cup butter or margarine
1 cup sugar
3 eggs
2 cups flour
2 teaspoons double-acting baking powder
1 teaspoon salt
1 cup plain yogurt
1 teaspoon baking soda
1 teaspoon vanilla extract
1 cup raisins

Topping:
    1 teaspoon ground cinnamon
    1/2 teaspoon grated nutmeg
    1/4 teaspoon ground cloves
    1/2 cup walnuts, chopped

confectioner's sugar, for topping

Preheat oven to 350°. Cream the butter and sugar in a large bowl. Add the eggs one at a time and beat until lemon-colored. Sift the flour with the baking powder and salt, then beat into the batter. Add the baking soda and vanilla to the yogurt; blend. Add the raisins and stir. Pour half the batter into a buttered 9-inch square baking pan. Top with half the mixture of cinnamon, nutmeg, cloves, and walnuts. Repeat this process.

Bake until the spicy cake springs back to the touch, about 40 to 45 minutes. Cool, sprinkle with confectioner's sugar, and cut into squares.

TRINIDADIAN PUMPKIN PIE

Called "Calabaza," this chilled, spicy pie is very versatile.
Serves 6.

8 oz. package cream cheese, softened
1/2 cup brown sugar
1 teaspoon ground cinnamon
1/2 teaspoon salt
1/2 teaspoon ground ginger
1/4 teaspoon ground cloves
1/4 teaspoon grated nutmeg
4 eggs
1 cup plain yogurt
1 cup cooked pumpkin, canned or fresh
1 teaspoon flavoring, vanilla or rum

9-inch prepared pie shell, unbaked
1 cup sour cream
2 Tablespoons brown sugar

Preheat oven to 350°. In a bowl, blend the softened cream cheese with the brown sugar, cinnamon, salt, ginger, cloves, and nutmeg. Add the eggs, one at a time; mix thoroughly. Combine the yogurt, pumpkin, and vanilla or rum flavoring. Add to the cream cheese-egg mixture, and turn into the prepared pie crust.

Bake for about 45 to 50 minutes, or until the pie tests done. Stir the brown sugar into the sour cream, spread on the hot pie, and return to the oven until the top is firmed, about 5 minutes. Chill until ready to serve.

TURKISH STRAWBERRY-PISTACHIO DESSERT

Luscious syrup-soaked sponges with a terrific topping.
Serves 8 to 10.

1 1/2 cups confectioner's sugar
1 cup plain yogurt
3 eggs
4 Tablespoons butter or margarine, melted
2 cups flour
1 Tablespoon lemon rind, grated
1 teaspoon double-acting baking powder
1/4 teaspoon baking soda

Syrup:
   1 cup sugar
   1 cup water
   1 Tablespoon lemon juice

Decoration:
   pistachio nuts, chopped
   whipped cream
   strawberries

Preheat oven to 300°. Beat the confectioner's sugar into the yogurt, then mix in the eggs, butter, flour, lemon rind, baking powder, and baking soda. Pour the batter into a greased 9-inch square pan and bake for 40 to 45 minutes, or until the cake tests done.

Meanwhile, make the *syrup* by bringing the sugar, water, and lemon juice to a boil in a small saucepan. Stirring constantly, boil for 10 minutes; keep hot.

When the cake is done, cut it into interesting shapes in the pan and pour the syrup over them until it is all absorbed. Sprinkle with pistachio nuts. When cool and ready to serve, top with whipped cream and strawberries.

YOGURT POPSICLE POSSIBILITIES

Handy frozen treats to encourage your creativity—make in ice trays or popsicle forms.
Makes 1 pint.

Frozen Fruit
>3/4 cup puréed or mashed fruit of your choice
>1/4 cup sugar
>2 cups plain yogurt
>1 teaspoon lemon juice

Sweeten the puréed or mashed fruit with the sugar. Blend in the yogurt and lemon juice. Put in containers, adding wooden sticks if you like, and freeze until firm, several hours.

Nutter-Butter Pops
>1/2 cup peanut butter, smooth or chunky
>1/2 cup honey
>2 cups plain yogurt
>chocolate syrup (optional)

Blend together the peanut butter, honey, and yogurt. If you want, put a teaspoon of chocolate syrup into each container before pouring in the peanut butter-yogurt mixture. Freeze.

Delicious Drinks

A man hath no better thing under the sun, than to eat,
and to drink, and to be merry.

–Ecclesiastes

Yogurt forms the base for delicious, nutritious, low-cal sipping, whether in simple shakes or spirited drinks.

The first case in point is *"Adults Only" Apricot Appetizer*, a healthy concoction into which those of legal age can toss some vodka. Another vodka specialty is *"Bloodies" for Brunch*, a variation on Bloody Marys. *Tahitian Lime Treats* uses either rum or gin.

A tradition since Biblical days is *Ancient Egyptian Cucumber Cooler*, topped with mint sprigs. *Pachadi*, made with tomatoes and hot spices, is popular in India.

Pakistani Pick-Me-Up is a thinned-down yogurt drink flavored with lemon, mint, chili powder, and salt. The Iranian version of thinned down yogurt, *Persian Abdug* is said to be an ideal cure for hangovers. And *Sweet Indian Lassi* can be made either spicy or sweet. *Turkish Yogurt Fizz* (Ayran) adds club soda.

Austrian Apricot Frost makes a very attractive and appetizing party punch. Or you might multiply the ingredients for *Peachy Cuban Eggnog*, which everyone appreciates.

The Chinese, who consider the peach a symbol of long life, share *Chinese Peach Blossoms*, which can be drunk as a late-day snack or early-morning "instant breakfast," as can *Polynesian Papaya Drink*. Another excellent elixir is *Hearty Health Drink*, which speaks for itself. *South American "Sorbete"* should also become a regular favorite.

Berries blend beautifully with yogurt, especially in shakes; try *Finnish Strawberry Shake*, then experiment with other berries. Ditto for *German Raspberry Refresher*. *Palestinian Orange-Banana*

Shake is best made with Jerusalem oranges–it's a sure winner with any age group.

Cordials and coffee are also yummy with yogurt. Be sure to try *Jamaican Tia Maria Café*, just the thing for an unusual après-ski sensation, or *Guzzling Grasshoppers*–which contain both creme de menthe and creme de cacao. Again, you can try out other combinations.

Bon appetit! And cheers!

DELICIOUS DRINKS

"ADULTS ONLY" APRICOT APPETIZER

The over-21 set can toss in some vodka.
Serves 2.

1 cup plain yogurt
1/2 cup nonfat dry milk
1 teaspoon vanilla extract
1 can (16 oz.) apricots, drained and chopped
1/4 cup club soda (or more, depending on your fizz quotient)
vodka

Put the yogurt, nonfat dry milk, vanilla, and apricots in an electric blender. Blend for 30 seconds at high speed. Add club soda to taste.

"Adults only" can add some vodka to their glasses.

ANCIENT EGYPTIAN CUCUMBER COOLER

Egyptians have enjoyed yogurt and cucumber drinks since Biblical days.
Makes a single serving.

1 cucumber, peeled, seeded, and finely chopped
1 cup plain yogurt
2 teaspoons lemon juice
1 teaspoon honey
1 teaspoon fresh dill, chopped
dash of celery salt

mint sprigs, for decoration

Blend together the finely chopped cucumber, yogurt, lemon juice, honey, dill, and celery salt (you might want to liquify the mixture in an electric blender). Pour into a tall glass and top with mint sprigs.

AUSTRIAN APRICOT FROST

Make a big batch of this and serve in a party punch bowl.
Serves 4 to 6.

2 cups apricot nectar
2 cups light white wine
4 Tablespoons lemon juice
1/4 cup light corn syrup
1 cup plain yogurt
1 cup sparkling water

strawberries, for decoration

Blend together the apricot nectar, white wine, lemon juice, corn syrup, and yogurt; add the sparkling water just before serving. Float fresh strawberries on top.

"BLOODIES" FOR BRUNCH

A yogurt variation on Bloody Marys.
Serves 6 to 8.

1 can (46 oz.) tomato juice
1/2 cup lemon juice
1 cup plain yogurt
2 Tablespoons Worcestershire sauce
2 to 3 drops hot pepper sauce
1 Tablespoon powdered sugar
1 teaspoon celery salt
1/2 teaspoon onion salt
1/4 teaspoon black pepper
2 Tablespoons fresh parsley, chopped
1 teaspoon dried basil
1 pint vodka

celery sticks, for decoration

Combine the tomato juice, lemon juice, yogurt, Worcestershire sauce, hot pepper sauce, powdered sugar, salts, pepper, parsley, and basil in a jug.

If making by the pitcher, either add the vodka or let individual brunchers judge their own amounts. Put a celery stick in each glass.

CHINESE PEACH BLOSSOMS

The Chinese consider the peach a symbol of long life.
2 servings.

2 peaches, pitted, peeled, and cut up
1 teaspoon lemon juice
1 cup plain yogurt
2 Tablespoons orange juice concentrate
brewer's yeast (optional)

Prepare the peaches, covering them with the lemon juice until ready.

Put the peaches, yogurt, and orange juice concentrate in a blender and whirl 30 seconds at medium speed. If you want to have this concoction for an "instant breakfast," you might add some brewer's yeast.

FINNISH STRAWBERRY SHAKE

Lingonberries, blueberries, or raspberries can also be used–we found all of them fabulous while in Finland.
A special after-school snack for 3 to 4 people.

2 cups strawberries, hulled and halved
1 cup plain yogurt
1 cup milk
2 scoops of ice cream (strawberry or vanilla–flavored)
2 Tablespoons sugar or honey
1 teaspoon vanilla

whole strawberries, for decoration

Combine the strawberries, yogurt, milk, ice cream, sugar or honey, and vanilla in an electric blender. Blend for 15 to 20 seconds at medium speed (or, longer and at a higher speed if you like it puréed.) Top each shake with an extra strawberry.

GERMAN RASPBERRY REFRESHER

Use any kind of berries, and freeze any leftovers into popsicles.
Makes 1 quart, which serves 4.

1 pint raspberry sherbet
2 cups plain yogurt
1/3 cup nonfat dry milk
1 teaspoon vanilla extract

raspberries, as garnish

Soften the sherbet slightly, then blend with the yogurt, nonfat dry milk, and vanilla for 30 seconds at medium speed in an electric blender.

Garnish each drink with whole fresh raspberries.

GUZZLING GRASSHOPPERS

These are like the cordial or the pie, not the insect.
Serves 4.

1/4 cup creme de menthe
2 Tablespoons creme de cacao
2 cups plain yogurt

Topping:
  whipped cream
  1/2 cup cream-filled chocolate cookies (about 6), crushed

Combine the creme de menthe, creme de cacao, and yogurt in an electric blender. Mix at high speed for 1 minute. Pour into cordial glasses and top with whipped cream and the crushed cookies.

Grasshoppers are great as after-dinner drinks.

HEARTY HEALTH DRINK

We have a relative in Vermont who swears by this daily drink.
2 to 3 servings.

1/2 cup fresh fruit (or 2 Tablespoons undiluted orange juice
concentrate)
2 Tablespoons blackstrap molasses
2 Tablespoons soy or safflower oil
1 teaspoon vanilla extract
1 cup plain yogurt
2 cups milk (regular, lowfat, or skim)
1/2 cup nonfat dry milk
1/3 cup brewer's yeast (for the really brave!)

In an electric blender combine the fruit, molasses, oil, vanilla, and yogurt. Blend at high speed for 1 minute. Add the milk, nonfat dry milk, and brewer's yeast and blend at medium speed for about 30 seconds more.

Gulp down quickly, thinking of your longevity.

JAMAICAN TIA MARIA CAFÉ

Just the thing for an unusual après-ski cup of coffee.
Makes 2 to 3 drinks.

1/2 cup coffee (regular or decaffeinated), cooled
1 cup milk
1 cup plain yogurt
4 Tablespoons sugar
2 Tablespoons Tia Maria

Topping:
    whipped cream
    ground cinnamon

Blend the coffee, milk, yogurt, sugar, and Tia Maria for 30 seconds at medium speed in an electric blender. Pour into mugs, top with whipped cream, and sprinkle lightly with cinnamon.

Note: You can also substitute creme de cacao, curaçao, or Grand Marnier for the Tia Maria.

PACHADI

A cool and nourishing spiced tomato drink popular in India.
Serves 2.

2 tomatoes, peeled, seeded, and finely diced
2 cups plain yogurt
4 Tablespoons ground coriander
1 teaspoon hot green chili pepper, minced
1/2 teaspoon mustard seed
1/2 teaspoon ground cumin
dash of salt

Put the tomatoes into a blender or food processor and purée. Stir in the yogurt, then add the coriander, hot chili pepper, mustard seed, cumin, and salt.

Note: Pachadi is actually so thick that you can eat it with a spoon.

PAKISTANI PICK-ME-UP

A refreshing lemon yogurt drink from Moslem country.
Serves 4.

1 quart water
2 cups plain yogurt
juice of 1 lemon
rind of 1 lemon, grated
1 teaspoon fresh mint, chopped
1/8 teaspoon chili powder
1/8 teaspoon salt

Combine all the ingredients: water, yogurt, juice and rind of 1 lemon, mint, chili powder, and salt in a jug.

Shake well and serve as is or with ice shavings.

PALESTINIAN ORANGE-BANANA SHAKE

Jerusalem oranges are best for this shake, but use whatever is available.
Snack for 1.

1 orange, peeled, seeded, and cut up
1 banana, peeled and cut into 1-inch chunks
1 cup plain yogurt
1/2 cup orange juice
2 Tablespoons honey
2 to 3 ice cubes or a scoop of ice cream (vanilla or fruit-flavored)

orange or banana slices, for garnish

Prepare the orange and banana; put into an electric blender with the yogurt, orange juice, honey, and ice cubes or ice cream. Mix at medium speed for 30 seconds.

Garnish with extra orange or banana slices.

PEACHY CUBAN EGGNOG

Cuban cuisine reflects fantastic Spanish and African influences.
Makes 2 tall drinks.

4 peaches, peeled and finely cut-up
1 can (6 oz.) orange juice concentrate
1/2 cup eggnog, commercial or homemade
1 cup plain yogurt

ground allspice, for garnish

Combine the cut-up peaches, orange juice concentrate, eggnog, and yogurt in an electric blender. Whirl at high speed until well blended, about 15 to 20 seconds.

Pour into glasses and sprinkle lightly with allspice.

PERSIAN ABDUG

Iranians claim that this drink is an ideal cure for hangovers.
A single serving.

1 cup ice water
1 cup plain yogurt
1/4 teaspoon salt

fresh sprigs of mint, for garnish

Whip together the ice water, yogurt, and salt, and serve with mint sprigs.

Note: India's "Spicy Lassi" is similar, but adds some ice cubes, 1/4 teaspoon cumin, and a pinch of cayenne pepper.

POLYNESIAN PAPAYA DRINK

The South Seas are famous for their fabulous fruits.
Makes 2 cups.

1 cup papaya juice
1 cup plain yogurt

2 teaspoons coconut, finely grated, for garnish

Blend together the papaya juice and yogurt in an electric blender. Pour into glasses and sprinkle with the grated coconut.

SOUTH AMERICAN "SORBETE"

This sorbete is a delicious fruit drink from A-B-C land.
Serves 1 to 2.

1 papaya, peeled, seeded, and chopped
1 banana, peeled and cut into 1-inch slices
1 can (11 oz.) mandarin oranges, undrained
4 Tablespoons sugar
1 cup plain yogurt

fruit slices, for garnish

Blend the papaya, banana, oranges, and sugar for 30 seconds at medium speed in an electric blender. Add the yogurt and blend for 30 seconds more.

Note: Try experimenting with other fruit-yogurt combinations.

SWEET INDIAN LASSI

People in India like their yogurt drinks either spicy or sweet.
Serves 6.

4 cups water
2 cups plain yogurt
2 Tablespoons honey
1/4 teaspoon grated nutmeg
pinch of cayenne pepper
dash of rosewater

In a large pitcher, stir the honey into the yogurt, then blend with the water, nutmeg, cayenne pepper, and rosewater. Serve chilled.

TAHITIAN LIME TREATS

Limes from Tahiti give great cause for celebration.
Serves 4.

1 can (6 oz.) frozen limeade concentrate, thawed
6 oz. water
6 oz. rum or gin
3 oz. club soda or quinine (tonic) water
1 cup plain yogurt
2 Tablespoons sugar, simple syrup, or sweetener to taste
1 banana, cut up (optional)

sherbet (lime is luscious!)

Combine the limeade, water, rum or gin, club soda or quinine water, yogurt, sugar, and banana in an electric blender. Whirl at high speed for 1 minute.

Top with a dollop of sherbet.

TURKISH YOGURT FIZZ

Club soda adds zip to this Turkish "Ayran."
Serves 4.

2 cups plain yogurt
2 cups club soda
pinch of garlic powder
dash of salt

Put the yogurt in a large pitcher. Stir in the club soda, garlic powder, and salt; blend.

Refrigerate until ready to serve.

Appendix A

What Yogurt Has Been Called
Around the World
(Latin: *Lactobacillus bulgaricus*)

Armenia–madzoon, mazoon, matzoon, mazun
Assyria–lebany
Balkans–tarho
Bulgaria–naja, kiselo mleko
Burma–tyre
Chile–skuta
Egypt–leben
Finland–plimoe, plimae
France–youghourt
Greece–tiaourti, oxygala, yaourti
Iceland–skyr
India–dahi, dahli, lassi, chass, matta
Iran–mast
Lapland–pauria, pauira
Lebanon–laban
Mongolia–koumiss
Norway–kaelder milk
Russia–prostokvasha, varenetz
Sardinia–gioddu
Sicily–mezzoradu
South Africa (Afrikaans)–joghurt
Sweden–filmjolk
Turkey–yoghurt, yogurt
United Kingdom and United States–yogurt

Appendix B

Countries Represented in *Yogurt, Yoghurt, Youghourt*

Africa
Albania
Armenia
Austria
Australia
Barbados
Bavaria
Belgium
Bulgaria
Canada
Ceylon
Chile
China
Czechoslovakia
Denmark
Egypt
Finland
France
Germany
Great Britain
Greece
Haiti
Hong Kong
Hungary
Iceland
India
Indonesia
Iran
Ireland

Israel
Italy
Jamaica
Kashmir
Lebanon
Malaysia
Mexico
Monaco
Morocco
Netherlands
New Zealand
Norway
Pakistan
Philippines
Poland
Polynesia
Portugal
Puerto Rico
Romania
Russia
Scotland
Singapore
Spain
Sweden
Switzerland
Thailand
Turkey
United States of America
Yugoslavia (former)

Index